AUTODESK INVENTOR™ FROM THE TOP

DANIEL T. BANACH
TRAVIS JONES

autodesk Press

Australia • Canada • Mexico • Singapore • Spain • United Kingdom • United States

Autodesk Inventor™: From the Top
Daniel T. Banach and Travis Jones

Autodesk Press Staff

Business Unit Director:
Alar Elken

Executive Editor:
Sandy Clark

Acquisitions Editor:
Jim DeVoe

Editorial Assistant:
Jasmine Hartman

Executive Marketing Manager:
Maura Theriault

Channel Manager:
Mary Johnson

Marketing Coordinator:
Karen Smith

Executive Production Manager:
Mary Ellen Black

Production Manager:
Larry Main

Production Editor:
Stacy Masucci

Art & Design Coordinator:
Mary Beth Vought

COPYRIGHT © 2002 Thomson Learning™. ⟨

Printed in Canada
2 3 4 5 XXX 05 04 03 02

For more information, contact Autodesk Press, 3 Columbia Circle, PO Box 15015, Albany, New York, 12212-15015.

Or find us on the World Wide Web at www.autodeskpress.com

All rights reserved. No part of this work covered by the copyright hereon may be reproduced or used in any form or by any means—graphic, electronic, or mechanical, including photocopying, recording, taping, Web distribution or information storage and retrieval systems—without written permission of the publisher.

To request permission to use material from this text contact us by
Tel: 1-800-730-2214
Fax: 1-800-730-2215
www.thomsonrights.com

Library of Congress Cataloging-in-Publication Data
Banach, Daiel T.
  Autodesk Inventor from the top / Daniel T. Banach.
    p.cm.
  ISBN 0-7668-4358-0
    1.Engineering graphics. 2.Engineering models--Data processing. 3.Autodesk inventor. I.Title

T353 .B17 2001
620'.0042'02855369--dc21
                          2001047829

Notice to the Reader
Publisher does not warrant or guarantee any of the products described herein or perform any independent analysis in connection with any of the product information contained herein. The publisher and author do not assume and expressly disclaim any obligation to obtain and include information other than that provided to it by the manufacturer.

The reader is expressly warned to consider and adopt all safety precautions that might be indicated by the activities described herein and to avoid all potential hazards. By following the instructions contained herein, the reader willingly assumes all risks in connection with such instructions.

The publisher and author make no representations or warranties of any kind, including but not limited to, the warranties of fitness for particular purpose or merchantability, nor are any such representations implied with respect to the material set forth herein, and the publisher and author take no responsibility with respect to such material. The publisher and author shall not be liable for any special, consequential, or exemplary damages resulting, in whole or in part, from the readers' use of, or reliance upon, this material.

Trademarks
Autodesk, the Autodesk logo, and AutoCAD are registered trademarks of Autodesk, Inc., in the USA and other countries. Thomson Learning is a trademark used under license. Online Companion is a trademark and Autodesk Press is an imprint of Thomson Learning. Thomson Learning uses "Autodesk Press" with permission from Autodesk, Inc., for certain purposes. All other trademarks, and/or product names are used solely for identification and belong to their respective holders.

CONTENTS

INTRODUCTION
AUTODESK INVENTOR: FROM THE TOP xv
PRODUCT BACKGROUND xv
PREREQUISITES FOR USING THIS BOOK xvi
BOOK'S INTENT xvi
SPECIAL SECTIONS xvi
BOOK NOTATIONS xvi
DESIGNING IN 3-D xvii
TERMS AND PHRASES xvii
TOOLBARS xviii
INTELLIMOUSE xviii
INCLUDED FILES ON THE CD xviii
PROJECT FILES FOR TUTORIALS AND EXERCISES xviii
THE TUTORIALS AND EXERCISES xix
OVERVIEW OF PART, ASSEMBLY, AND DRAWING VIEW CREATION xix
ACKNOWLEDGMENTS xix
DEDICATIONS xx

CHAPTER 1 Getting Started, Sketching, Constraining, and Dimensioning
GETTING STARTED 1
    Getting Started 2
    New 2
    Open 2
    Projects 3
    Project Files 5
TUTORIAL 1.1—PROJECTS 6
    File Types 6
APPLICATION OPTIONS 7
    General Tab 7
    File Tab 9
    Colors Tab 10
    Display Tab 11
    Hardware Tab 14
DESIGN SUPPORT SYSTEM 14
USER INTERFACE 15
    Toolbars and Panel Bar 16

COMMAND ENTRY 17
    Toolbars and Panel Bar 17
    Shortcut Menus 17
    Windows Shortcuts..17
    Hot Keys 18
    Undo and Redo 19
SKETCHING AND PART APPLICATION OPTIONS 19
    Sketch Options 19
    Part Options 21
UNITS 22
TEMPLATES 23
CREATING A PART 24
    Sketches and Default Planes 24
    New Sketch 25
SKETCH THE OUTLINE OF THE PART: STEP 1 26
    Sketching Overview...26
    Sketching Tools 27
    Using the Sketch Tools 29
TUTORIAL 1.2—SKETCHING 32
CONSTRAINING THE SKETCH: STEP 2 32
    Dragging a Sketch 33
    Constraint Types 33
    Showing and Deleting Constraints...34
    Adding Constraints 35
ADDING DIMENSIONS: STEP 3 36
    General Dimensioning 36
    Dimensioning Lines 37
    Dimensioning an Angle 37
    Dimensioning Arcs and Circles 37
    Diametric Dimensions 37
    Dimensioning to a Quadrant 38
ENTERING AND EDITING DIMENSION VALUES 38
OVER-CONSTRAINED SKETCHES 38
TUTORIAL 1.3—CONSTRAINING AND DIMENSIONING 39
PRACTICE EXERCISES 41
EXERCISE 1.1—BRACKET 41
EXERCISE 1.2—CONNECTING ROD 42
REVIEW QUESTIONS..43

CHAPTER 2   Viewing, Extruding, Revolving, Editing Features, Sketched Features, and Boolean Operations

VIEWING A MODEL FROM DIFFERENT VIEWPOINTS 44
    Isometric View 45
    Camera Views..45

View Tools	45
Display Options	48

TUTORIAL 2.1—VIEWING A MODEL 49
UNDERSTANDING WHAT A FEATURE IS 50
USING THE BROWSER FOR CREATING AND EDITING 50
SWITCHING ENVIRONMENTS 51
FEATURES TOOLS 52
EXTRUDING THE SKETCH 53

Shape	54
Operation	54
Extents	55

TUTORIAL 2.2—EXTRUDING A SKETCH 56
REVOLVING THE SKETCH 59

Shape	59
Operation	59
Extents	60

CENTERLINES AND DIAMETRIC DIMENSIONS 60
TUTORIAL 2.3—REVOLVING A SKETCH 61
EDITING A FEATURE (FEATURE EDITING) 64

Failed Features	65
Renaming Features and Sketches	65
Feature Color	66
Deleting a Feature	66

TUTORIAL 2.4—EDITING FEATURES 67
EDITING A FEATURE'S SKETCH 68
TUTORIAL 2.5—EDITING A FEATURE'S SKETCH 69
SKETCHED FEATURES 70
ASSIGNING A PLANE TO THE ACTIVE SKETCH 70

Face Cycling	71

TUTORIAL 2.6—SKETCH PLANES 72
TUTORIAL 2.7—SKETCH PLANES 73
CUT, JOIN, AND INTERSECT OPERATION 73

Operation	74
Termination	75

TUTORIAL 2.8—OPERATIONS AND TERMINATIONS 75
PRACTICE EXERCISES 85
EXERCISE 2.1—BRACKET 85
EXERCISE 2.2—CONNECTING ROD 85
EXERCISE 2.3—CYLINDER 86
REVIEW QUESTIONS 87

CHAPTER 3 Placed Features

FILLETS 89

Constant Tab	90
Variable Tab	91

Setbacks 92
TUTORIAL 3.1—CREATING FILLETS 93
CHAMFERS 96
    Method 97
    Edge and Face 98
    Distance and Angle 98
    Edge Chain and Setback 98
TUTORIAL 3.2—CREATING CHAMFERS 98
HOLES 101
    Hole Centers 104
TUTORIAL 3.3—CREATING HOLES 105
THREADS 108
    Location Tab 108
    Specification Tab 109
TUTORIAL 3.4—CREATING THREADS 109
SHELLING 111
    Remove Faces 112
    Thickness 112
    Direction 113
    Unique Face Thickness 113
TUTORIAL 3.5—SHELLING A PART 113
FACE DRAFT 115
    Pull Direction 115
    Faces 115
    Draft Angle 116
TUTORIAL 3.6—APPLYING FACE DRAFT 116
CREATING A WORK AXIS 119
CREATING WORK POINTS 119
TUTORIAL 3.7—CREATING A WORK AXIS AND WORK POINTS 120
CREATING WORK PLANES 121
    Feature Visibility 123
TUTORIAL 3.8—CREATING WORK PLANES 123
TUTORIAL 3.9—CREATING WORK PLANES 126
TUTORIAL 3.10—CREATING WORK PLANES 128
PATTERNS 130
    Rectangular Patterns 130
    Circular Patterns 132
TUTORIAL 3.11—RECTANGULAR PATTERNS 133
TUTORIAL 3.12—CIRCULAR PATTERNS 134
PRACTICE EXERCISES 136
EXERCISE 3.1—DRAIN PLATE 136
EXERCISE 3.2—CYLINDER 136
REVIEW QUESTIONS 138

CHAPTER 4 Creating and Editing Drawing Views

- DRAWING OPTIONS, CREATING A DRAWING, AND DRAWING TOOLS 140
  - Drawing Options 140
  - Creating a Drawing 141
  - Drawing Tools 142
- DRAWING SHEETS PREPARATION 142
- BORDER CREATION 143
- TITLE BLOCK CREATION 143
  - Default Title Block 143
  - Creating a New Title Block 144
- TUTORIAL 4.1—SHEETS, BORDERS, AND TITLE BLOCKS 145
- SETTING UP DRAFTING AND DIMENSION STANDARDS 148
  - Drafting Standards 148
  - Dimension Styles 149
- TEMPLATE 151
- CREATING DRAWING VIEWS 151
  - Base Views 152
  - Projected Views 154
  - Auxiliary View 155
  - Section View 155
  - Detail View 156
  - Broken View 157
- DRAFT VIEWS 158
- TUTORIAL 4.2—DRAWING VIEWS 159
- EDITING DRAWING VIEWS 164
  - Moving Drawing Views 164
  - Edit Drawing View Properties 164
  - Deleting Drawing Views 166
- DIMENSIONS 166
  - Dimension Visibility 166
  - Dimensions Value and Appearance 167
  - Drawing (Reference) Dimensions 167
  - Auto Baseline Dimensions 168
  - Ordinate Dimensions 169
  - Moving Dimensions 172
  - Hole Tables 172
- TUTORIAL 4.3—EDITING DRAWING VIEWS AND DIMENSIONING 173
- ANNOTATIONS 175
  - Centerlines 175
  - Surface Textures, Weld Symbols, and Feature Control Frames 177
- SKETCHED SYMBOLS 177
  - Text 178
- HOLE AND THREAD NOTES 178

PARTS LIST 178
    Parts List Operations (Top Icons) 180
    Spreadsheet View 181
    More Sections of the Parts List Dialog 181
CREATING AN AUTOCAD 2-D DRAWING FROM INVENTOR
    DRAWING VIEWS 181
TUTORIAL 4.4—ANNOTATIONS 182
    Review-Sequence for Creating Drawing Views 184
PRACTICE EXERCISES 185
EXERCISE 4.1—DRAIN PLATE 185
EXERCISE 4.2—CONNECTOR 186
REVIEW QUESTIONS 187

CHAPTER 5 Assemblies

CREATING ASSEMBLIES 189
ASSEMBLY OPTIONS 190
    Part Feature Adaptivity 191
    In-Place Features 191
    Shared Content Link 192
BOTTOM UP APPROACH 192
TOP DOWN APPROACH 193
    Occurrences 193
    Multiple Documents and Drag and Drop Components 194
    Active Component 194
    Open and Edit 194
GROUNDED COMPONENTS 194
SUBASSEMBLIES 195
REORDER AND RESTRUCTURE 196
TUTORIAL 5.1—BOTTOM UP ASSEMBLY TECHNIQUE 196
TUTORIAL 5.2—TOP DOWN ASSEMBLY TECHNIQUE 197
ASSEMBLY CONSTRAINTS 201
    Types of Constraints 202
    Assembly Tab 203
    Motion Tab 203
    Transitional Tab 204
    Constraint Types 204
    Applying Assembly Constraints 207
EDITING ASSEMBLY CONSTRAINTS 208
DRIVING CONSTRAINTS 209
INTERFERENCE CHECKING 211
TUTORIAL 5.3—BOTTOM UP ASSEMBLY CONTINUED 212
TUTORIAL 5.4—TOP DOWN ASSEMBLY CONTINUED 215
iMATES 218
REPLACING COMPONENTS 220

PATTERNED COMPONENTS 220
    Component Pattern 221
    Assembly Pattern 222
TUTORIAL 5.5—iMATES, REPLACING, AND PATTERN COMPONENTS 222
ASSEMBLY WORK FEATURES..226
ADAPTIVITY..226
    Sketches..228
    Features..229
    Subassembly..229
    Adapting the Sketch or Feature 230
TUTORIAL 5.6—ADAPTIVITY 230
ENABLED COMPONENTS 234
DESIGN VIEWS 234
PRESENTATION FILES..235
    Creating Presentation Views 237
    Tweak Components..238
    Animation 240
TUTORIAL 5.7—PRESENTATION VIEWS 242
CREATING DRAWING VIEWS FROM ASSEMBLIES AND PRESENTATION
    FILES..245
CREATING BALLOONS...246
    Parts List Item Numbering Dialog Box Options.....................................247
PARTS LIST 248
TUTORIAL 5.8—ASSEMBLY DRAWING VIEWS 249
PRACTICE EXERCISES 252
EXERCISE 5.1—ADAPTIVE ASSEMBLY 252
REVIEW QUESTIONS..255

CHAPTER 6 Advanced Dimensioning, Constraining, and Sketching Techniques

CONSTRUCTION GEOMETRY...257
ELLIPSES...257
PATTERN SKETCHES 258
SHARED SKETCHES 259
MIRROR SKETCHES AND SYMMETRY CONSTRAINT 260
TUTORIAL 6.1—ELLIPSES, PATTERN SKETCHES, SHARED SKETCHES,
    AND THE MIRROR TOOL 261
SLICE GRAPHICS 263
PROJECT EDGES..264
SKETCH ON ANOTHER PART'S FACE 265
TUTORIAL 6.2—PROJECTING EDGES AND SKETCHING ON ANOTHER
    PART'S FACE 266
AUTO DIMENSION..270

DIMENSION DISPLAY, RELATIONSHIPS, AND EQUATIONS 271
    Dimension Display 271
    Dimension Relationships 272
    Equations 272
PARAMETERS 272
PARAMETERS LINKED TO A SPREADSHEET 275
TUTORIAL 6.3—AUTO DIMENSION, RELATIONSHIPS, PARAMETERS, AND SPREADSHEETS 277
iPARTS 282
    Creating iParts 283
    iPart Placement 286
TUTORIAL 6.4—iPARTS 288
2-D DESIGN LAYOUT 292
TUTORIAL 6.5—2-D DESIGN LAYOUT 292
PRACTICE EXERCISES 294
EXERCISE 6.1—iPART-RECTANGULAR STEEL TUBING 294
EXERCISE 6.2—2-D LAYOUT 295
REVIEW QUESTIONS 296

CHAPTER 7 Advanced Modeling Techniques

SWEEP 298
    Shape 298
    Operation 299
TUTORIAL 7.1—SWEEP FEATURES 299
COIL FEATURES 300
    Coil Shape Tab 300
    Operation 301
    Coil Size Tab 302
    Coil Ends Tab 302
TUTORIAL 7.2—COIL FEATURES 303
LOFT FEATURES 308
    Sections 308
    Operation 309
    Shape Control 309
    More Options 310
TUTORIAL 7.3—LOFT FEATURES 310
PART SPLIT 312
    Method 312
    Faces 313
TUTORIAL 7.4—SPLITTING A PART 314
COPY FEATURES 316
TUTORIAL 7.5—COPY/PASTE FEATURE 317
REORDER FEATURES 322
MIRROR FEATURES 322
TUTORIAL 7.6—MIRRORING ABOUT A PLANE 323

FEATURE SUPPRESSION	325
USING SURFACES	326
RIBS AND WEBS	326
Shape	327
Thickness	327
Extents	327
RIB NETWORKS	328
TUTORIAL 7.7—CREATING RIBS AND WEBS	328
iFEATURES	332
Create iFeatures	332
TUTORIAL 7.8—CREATE iFEATURES	334
Insert iFeatures	336
TUTORIAL 7.9—INSERTING iFEATURES	337
iFeatures Options	339
DERIVED PARTS	340
Derived Part Symbols	341
Derived Part Dialog	342
Derived Assembly Symbols	343
Derived Assembly Dialog	343
TUTORIAL 7.10—CREATE A DERIVED PART	344
TUTORIAL 7.11—CREATE A DERIVED ASSEMBLY	346
FILE PROPERTIES	347
Summary	347
Project	347
Status	348
Custom	348
Save	349
Physical	349
REVIEW QUESTIONS	349

CHAPTER 8 Additional Functionality

ENGINEERS NOTEBOOK	350
Adding Notes to a Model	351
Adding Other Content	353
Managing Notes	353
Note Display	353
SHEET METAL PARTS	353
Sheet Metal Design Methods	354
Creating a Sheet Metal Part	354
SHEET METAL TOOLS	354
Sheet Metal Styles	355
Face	357
Contour Flange	359
Flange	361
Hem	361

Fold	363
Bend	364
Corner Seam	364
Cut	366
Corner Round	367
Corner Chamfer	367
PunchTool	368
Flat Pattern	369
Common Tools	370
TUTORIAL 8.1—SHEET METAL PART	**370**
PARTS LIBRARY	**383**
RedSpark and i-Drop	383
Library Setup	383
Place Content	384
TUTORIAL 8.2—FASTENERS AND REDSPARK PARTS LIBRARY	**385**
DESIGN ASSISTANT	**385**
Design Assistant Modes	386
DWG (AUTOCAD AND MECHANICAL DESKTOP) WIZARDS	**388**
Overview	388
Import	389
Export	395
TUTORIAL 8.3—DWG IMPORT	**397**
IMPORTING OTHER FILE TYPES	**399**
REVIEW QUESTIONS	**400**
INDEX	**401**

AUTODESK INVENTOR: FROM THE TOP

Welcome! If you are new to Autodesk Inventor or have experience with Inventor, you will find this book informative as a learning tool and as a reference guide.

The chapters in this book follow the order in which you will create your own models and drawings. Each chapter introduces a set of topics and then takes you through a basic, step-by-step example. Each chapter builds on the material learned in the previous chapter(s). At the end of most chapters you will find practice exercises for you to complete on your own. They are based on real world parts used in different disciplines of design.

PRODUCT BACKGROUND

Inventor was written by Autodesk to do mechanical design. Inventor is a 3-D feature-based parametric solid modeler that goes beyond parametrics by allowing parts to be adaptive. Adaptivity allows relationships to be built between parts that help define the parts shape. Once a 3-D part or assembly is created, drawing views can be automatically generated.

Inventor is a stand-alone application that consists of the following main modules:

Part Modeling: Feature-based parametric and adaptive modeler.

Sheet Metal Modeling: Create parametric and/or adaptive sheet metal parts and flat patterns.

Assembly Modeling: Manage and constrain parts in an assembly.

Fastener Library: Insert standard nuts and bolts into your assembly.

Presentation Creation: Create exploded views and/or animations of an assembly that show how the parts are assembled or disassembled.

Drawing Creation: Generate 2-D drawing views with dimensions and annotations.

Engineer's Notebook: Communicate your thoughts by creating graphical notes that are stored with a part or assembly.

Design Assistant: Manage part and assembly properties and relationships.

PREREQUISITES FOR USING THIS BOOK

This book assumes that you are new to Inventor with no CAD experience but do have Windows experience. If you have prior 3-D CAD experience you will also find this book easy to follow and learn the concepts quickly.

BOOK'S INTENT

The intent of this book is to focus on part modeling, assembly modeling, and drawing view creation in an applied, hands-on environment. With software changing at a rapid pace, it is the intent of this book to give a general overview on how to get the most out of the software in a short time. The topics that were selected for this book will allow you to create parts, assemblies, and annotate them. Not every operation or option of every tool will be covered. After completing an operation in the book, you are encouraged to investigate the other options that were not covered. Utilize Inventor's online help system to get more information about certain tools. This book guides you through the process of generating 3-D parametric and adaptive parts, assembling parts together and producing 2-D views from them. Both bottom-up and top-down assembly modeling techniques will be covered.

Each chapter will be broken into specific subjects, which are introduced and then followed by a short tutorial. When a new tool is introduced, there is a figure showing the icon in the specific toolbar as well as how to access the tool through right click menus. Read through each subject and then complete its tutorial while at the computer. At the end of each chapter, there are exercises you can complete on your own and review questions to reinforce the topics covered in that chapter.

SPECIAL SECTIONS

You will find sections marked:

 Note: Here you will find information that points out specific areas that will help you learn Inventor.

 Tip: Here you will find information that will assist you in generating better models.

BOOK NOTATIONS

- Autodesk Inventor and Inventor both refer to the product Autodesk Inventor.
- ENTER refers to Enter on the keyboard or a right mouse click.
- Numbers in quotation marks are numbers that need to be typed in.
- Component part and model refer to an Inventor part.
- Select refers to a click by the left mouse button.

- Pick and choose refer to selections from a menu.
- RMC means to right mouse click.
- Inventor's browser and Browser both refer to the Inventor browser, where the history of the file is shown.
- Sketch refers to lines, arcs, circles, and splines drawn to define an outline shape of a part or feature.

DESIGNING IN 3-D

If you are new to creating 3-D parts, take time to evaluate what you are going to model and how you are going to approach it. When I evaluate a part, I look for the main basic shape. Is it flat or cylindrical in shape? Depending on the shape, I will take a different approach to the model. I try to start with a flat face if possible; from a flat face it is easier to add other features. If the model is cylindrical in shape, I look for the main profile or shape of the part and revolve or extrude that profile. After the main body is Created, work on the other features, looking for how this shape will connect to the first part. Think about 3-D modeling like working with building blocks, each block sitting on another block, but remember that material can also be removed from the original solid.

TERMS AND PHRASES

To help you to better understand Inventor, a few of the terms and phrases that will be used in the book are explained below.

Parametric Modeling: Parametric modeling is the ability to drive the size of the geometry by dimensions, sometimes referred to as "Dimension Driven Design." For example, if you want to increase the length of a plate from 5″ to 6″, change the 5″ dimension to 6″ and the geometry will update. Think of it as the geometry being along for a ride, driven by the dimensions. This is opposite of associative dimensions: as lines, arcs, and circles are drawn, they are created to the exact length or size; when they are dimensioned, the dimension reflects the exact value of the geometry. If you want to change the size of the geometry, you stretch the geometry and the dimension automatically gets updated. Think of this as the dimension being along for a ride, driven by the geometry.

Feature-Based Parametric Modeling: Feature-based means that as you create your model, each hole, fillet, chamfer, extrusion, and so on is an independent feature that can have its dimensional values be changed without having to re-create them.

Adaptivity: Adaptivity allows parts to have a physical relationship between each other; for example, if a plate is created with a dimensioned cutout and another plate is created with the same cutout but NO dimensions are created. The second cutout can be constrained to match the size of the dimensioned cutout. When the dimensions change on the first cutout, the second cutout will update or adapt to show the change.

Bi-directional Associativity: The model and the drawing views are linked. If the model changes, the drawing views will automatically update. And if the dimensions in a drawing view change, the model is updated and the drawing views are updated based on the updated part.

Bottom-Up Assembly: Refers to an assembly whose parts were created in individual part files outside of the assembly and then referenced into the assembly.

Top-Down Assembly: Refers to an assembly whose parts were created while working in the assembly file.

TOOLBARS

In this book, the toolbars will be shown in the default orientation. The toolbars' location is a personal choice and can be placed and oriented anywhere you want. If you prefer to have the toolbars take up less space, you can turn on the expert mode (removes text from the icon) by right clicking in the Panel Bar, and check the Expert line from the pop-up menu.

INTELLIMOUSE

Autodesk Inventor takes full advantage of a wheel mouse (mouse with a center wheel between the buttons). If you are using a wheel mouse you can zoom in and out by scrolling the wheel. To pan, hold the wheel and pan in the direction you want. You can use a regular two or three button mouse, but you will not get the functionality of a wheel mouse.

INCLUDED FILES ON THE CD

Each book ships with a CD that contains files that will be needed to complete the tutorials and exercises. All of the needed files will be stored in the directory \Inventor Book\.

PROJECT FILES FOR TUTORIALS AND EXERCISES

The \Inventor Book\ directory contains the files for the tutorials and exercises. If the files are moved to a different directory, the project file will need to be edited to reflect the new directory structure. Creating and editing projects will be covered in Chapter 1.

If the files are copied from the CD to your hard drive, the files will be read-only and will need to have the read-only property removed from them.

THE TUTORIALS AND EXERCISES

The best way to learn Inventor is to practice the tutorials and exercises on the computer. As you go through the book you will find some of the exercises have been started for you; they can be opened as noted in each exercise. As you go through each exercise, save the files as noted because they may be required in a future tutorial or exercise.

For clarity, some figures will be shown with lines hidden and the text larger.

There are many ways to complete a part or assembly when creating 3-D models. After completing the exercises as noted, feel free to experiment with different methods.

OVERVIEW OF PART, ASSEMBLY, AND DRAWING VIEW CREATION

To get a better idea of the process that you will go through to create a part, assemble it to other part(s), and create 2-D views, refer to the outline below. This is intended as an overview only.

- Draw a sketch.
- Add constraints and dimensions.
- Turn the sketch into a 3-D part by extruding, revolving, or sweeping.
- Add sketched and placed features.
- Assemble parts together.
- Create drawing views with annotations.

ACKNOWLEDGMENTS

Dan Banach would like to thank Travis Jones for helping to both write and technical edit this book. His professionalism was greatly appreciated.

Travis Jones thanks Dan Banach for presenting the opportunity to work on this project with him. His knowledge, understanding, and work ethic are incredible. Thanks also go to Neil Munro for his knowledge and input on Autodesk Inventor and especially its Sheet Metal capabilities.

Special thanks go to the following instructors who reviewed the content:

Steve Brown, College of the Redwoods, Eureka, CA

Tom Singer, Sinclair Community College, Dayton, OH

DEDICATIONS

I dedicate this book to my wife, Cammi, and my children, Allison and Jonathan. Without their love, understanding, and encouragement this book would not have been possible.

Daniel T. Banach

To my kids, Jake and Dakota, who supported this task and willingly sacrificed their time. Most especially to my wife, Kelli, for her love, encouragement, and support. They are the ones who helped make this happen.

Travis Jones

CHAPTER 1

Getting Started, Sketching, Constraining, and Dimensioning

The first step in creating 3-D parametric parts and assembling them together is to start by creating a 2-D sketch that will be transformed into a 3-D part. Most 3-D parts in Autodesk Inventor will start from a 2-D sketch. Before getting into the creation of 2-D sketches we will look at Inventor's user interface, application options, starting tools (commands) and then we will cover the three steps in generating a 2-D parametric sketch. After that we will sketch a rough 2-D outline of the part, apply geometric constraints, and then add parametric dimensions.

CHAPTER OBJECTIVES

After completing this chapter, you will be able to:

- understand the different file types that Autodesk Inventor uses.
- understand what a project file is used for.
- understand Autodesk Inventor's user interface.
- describe Autodesk Inventor's sketch and part options.
- sketch an outline of a part.
- understand and create geometric constraints.
- dimension a sketch.
- change a dimensions values in a sketch.

GETTING STARTED

After starting Autodesk Inventor you will notice that your screen will look similar to Figure 1.1. On the left side of the screen you will have four options of what to do: Getting Started, New, Open, and Projects.

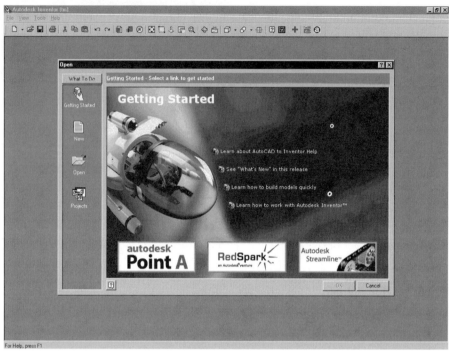

Figure 1.1

GETTING STARTED

Under this option you will find information about Autodesk Inventor tools; how to use Autodesk Inventor to create parts and assemblies and annotate them; and to find out what is new in the current release. This area can be very useful to help reinforce what you are learning in this book.

NEW

After selecting the new icon you will be asked to create a new part, assembly, presentation file, sheet metal part or a drawing as shown in Figure 1.2. By selecting the tabs you can create new files based on other drafting standards. New tabs can be added by creating a subdirectory under the Inventor "version number"\Templates directory and placing files into it. Templates will be covered in more detail later in this chapter.

OPEN

After selecting the Open icon you will be prompted to choose an Autodesk Inventor file to open. The directory that appears by default is set in the current project file. You can open files from other directories, but it is not recommended since part, drawing, and assembly relationships may not be resolved and could lead to difficulties when opening an assembly.

Figure 1.2

PROJECTS

Projects are text files that direct where Inventor should look for files (search paths) that are needed for a given project. Without these search paths, Inventor would not be able to locate all the needed files when an assembly, drawing, or presentation file is opened. There is no limit to the number of projects that can be set up, but only one project can be active at any given time. After the project is created, the information is saved to the current workspace, and a shortcut to the project file is saved to your project folder. The project folder can be set within the Tools pull-down menu, Application Options selection, or from the File tab. Before creating a project ALL Autodesk Inventor files must be closed. When Inventor is opening a file, it searches for the file in the directories listed in the project.

Most small projects will usually store all files in a single directory; this directory will be specified in the workspace parameter. If needed you can specify local or library search paths where standard parts are located.

After selecting the Project Icon in the Getting Started area, a dialog box will appear as shown in Figure 1.3. The dialog box is divided into two areas. The top portion lists the project files that are available. To make a project active, double click on its name. Once the project is active the information about that project will appear on the bottom half of the dialog box.

- **Included File:** If you are working with a group and need to look for files that are in another project, you can add a second project file here. If you are working on a project whose files are in a single project, leave this option blank. The Included file can contain only one additional project file.

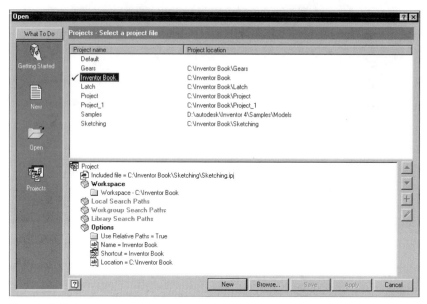

Figure 1.3

- **Workspace:** This is the location that your files will be saved to and the default location that files will be opened from. If all the files for a project are in a single directory, this entry is the only one needed for the project. The Workspace path can only point to one directory.

- **Local Search Paths:** If your project needs files that are not in the workspace and are on your local computer, add its path here. There can be multiple local search paths.

- **Workgroup Search Paths:** If your project needs files that are not in the workspace and are located on a network, add its path here. There can be multiple workgroup search paths.

- **Library Search Paths:** If your project needs files that are located in a standard or custom library, add the path here. There can be multiple library search paths.

- **Options:** Sets the defaults for the project.

    Use Relative Paths: Allows you to use relative paths when specifying directories within the project file. Relative Paths are designated by adding a '.' to the beginning of the path. The '.' tells Inventor to start the directory path in the same location where the project file is being stored.

    Name: Shows the name of the current project.

Shortcut: Shows the name of the shortcut that is displayed in the list of projects. To edit its name, right click on its name and select Edit from the context menu. Then type in a new name.

Location: Shows the location where the project file is being stored.

PROJECT FILES

Creating a Project

To create a new project, follow these steps.

1. Select the Project Icon from the What To Do area of the startup menu. If you are already working in Autodesk Inventor from the File pull down menu, select Projects (any open Inventor files need to be closed). If Autodesk Inventor is not running, you can create a new project by picking on the Microsoft Windows® Start menu and then select Programs>Inventor "version number">Tools>Project Editor.

2. Click the New button in the Projects dialog box.

3. You will be prompted by the Project wizard for the following: name, type of project, location, and libraries for the new project. If new directories are needed they will be created.

4. After the project wizard is finished, you can make the project current by double clicking on the project name in the top half of the dialog box.

Editing a Project

To edit project locations, follow these steps.

1. Select the Project Icon from the What To Do area of the startup menu. If you are already working in Autodesk Inventor from the File pull down menu, select Projects (any open Inventor files need to be closed). If Autodesk Inventor is not running you can create a new project by picking on the Microsoft Windows® Start menu and then select Programs>Inventor>Tools>Project Editor.

2. In the top portion of the Projects dialog box, double click on the name of the project to make the project active.

3. In the lower portion of the Projects dialog box, right mouse click on the section to edit and then from the context menu select the option that you want to do. Alternatively, you can select the area to edit then click the Add or Edit icon from the right side of the dialog box.

4. To edit the order that the directories are searched, select the directory path to move, then click the Up or Down arrow from the right side of the dialog box.

Tip: Use names for titles and directories that are descriptive. Do not create parts without first setting an active project to define the location for the files.

TUTORIAL 1.1—PROJECTS

1. Start Autodesk Inventor.
2. Create a new Personal Project named "Sketching." See Figure 1.4 for how the settings in the project wizard dialog box should look.
3. After creating the "Inventor Book" Project, make it the current project by double clicking on its name, and a check mark will appear before its name as shown in Figure 1.3.

This project will be used for all the tutorials and exercises for this book.

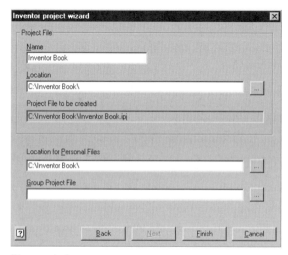

Figure 1.4

FILE TYPES

Autodesk Inventor uses a variety of file extensions for different types of files. Below is a listing of the file types that Autodesk Inventor can create and what they are used for.

- A "Part" file contains only one part and can be either 2-D or 3-D. A part file has an extension of IPT.
- An "Assembly" file can consist of a single part, multiple parts or subassemblies. The parts themselves are saved to their own part file and are referenced (linked) into the assembly file. The extension of an assembly file is IAM. Chapter 5 will cover assemblies.
- A "Presentation" file shows parts of an assembly being exploded in different states. The presentation file is linked to an assembly like parts are linked to an assembly. The presentation file can also be animated showing the parts being

assembled and/or disassembled. The animation can be saved as an AVI file. The extension of a presentation file is IPN. Chapter 5 will cover creating presentation files.

- A "Sheet Metal" file is a part file that has the sheet metal application loaded and has specific functionality for creating sheet metal parts and flat patterns. You can also create a sheet metal part while in a regular part; the sheet metal application will just need to be loaded manually. Chapter 8 will cover sheet metal.

- A "Drawing" file can contain 2-D projected drawing views of parts, assemblies, and or presentation files. Dimensions and annotation can also be added. Like the assembly and presentation files, the parts and assemblies are linked to the drawing file. The extension of a drawing file is IDW. Chapter 4 will cover drawing views.

- A "Project" file is an ASCII based text file that is used to direct Autodesk Inventor where to look for files. A project file has an extension of IPJ.

- iFeatures are files that can contain complete parts, 3-D features, or 2-D sketches that can be inserted into a part file where needed. Size limits and ranges can be placed on these iFeatures to further enhance their functionality. iFeature files have an extension of IDE. Chapter 7 will cover iFeatures.

APPLICATION OPTIONS

Autodesk Inventor can be customized to your preferences. Under the Tools pull down menu select Application Options and the Options dialog box will appear that looks like Figure 1.5. Under each of the tabs you will be able to control how Autodesk Inventor acts. Since there are numerous options, the options that affect certain areas of the book will be covered when that section is introduced. Here we will look at the General, File, Colors, Display, and Hardware tabs. For more information on the options utilize the online help.

GENERAL TAB

Under the General tab you have the following options.

- **Maximum Size of Undo File (Mb):** Select the maximum file size to be used for the Undo option of Autodesk Inventor. These undo files are temporary. When working with files that are large in size you will want to increase this file size. Autodesk recommends changing the file size in increments of 4-megabyte increments.

- **Number of Versions to Keep:** When a file is saved it stores the number of versions that are set here. The number of versions that can be stored in a file are 1 to 10. When the maximum number of versions is attained, the oldest version is deleted from the file. When opening a file select the option button to pick the version to open. When a file is opened, only the current version is loaded in memory.

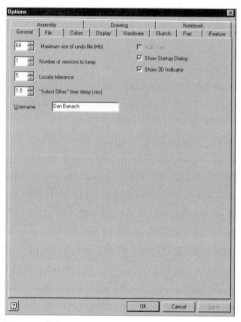

Figure 1.5

- **Locate Tolerance:** In pixels enter a number from 1 to 10; this is the distance you can be away from an object when you try to select it.

- **"Select Other" Time Delay (sec):** Set the number of seconds that will have to expire before you are allowed to select another feature or part.

- **Username:** Type in the name that will automatically appear when the Username is needed; an example would be the username is that used to populate the Author variable in the title block.

- **Defer Update:** When selected, updates to an assembly file are deferred until the assembly is updated. When the check box is not checked, the assembly is automatically updated after a part is edited.

- **Multi User:** When checked, Autodesk Inventor will reserve a part that you are editing to notify other users that you are working on the part. When deselected, Autodesk Inventor will not reserve the part.

- **Show Startup Dialog:** If you want to see the Startup dialog box, check this box.

- **Show 3-D Indicator:** When selected a 3-D indicator that represents the X, Y and Z axis will appear on the lower left corner of the screen. The axes are represented with colored arrows, X = red, Y = green and Z = blue.

FILE TAB

Under the File tab (Figure 1.6) you have the following options.

- **Undo:** Specify the directory where Inventor's temporary files will be stored.
- **Templates:** Select the directory where the template files are located.
- **Project Folder:** Set the location for Autodesk Inventor to look for project files. When new projects are created they are added to this directory.
- **Workgroup Design Data:** Set the location for external tables that Autodesk Inventor needs such as thread tables.
- **Default VBA Project:** Set the location for Autodesk Inventor to look for VBA files.
- **Transcripting:** In this area you can turn transcripting on and off and set the file location. Transcripting records every operation that you do and can be replayed to recreate the file. Only use transcripting if you are having problems with a part or you want to show someone how to do a certain task. The file size can get very large.

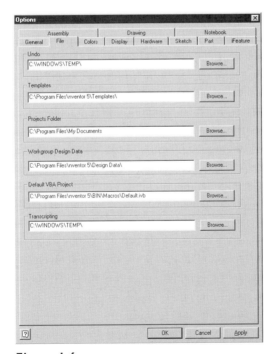

Figure 1.6

COLORS TAB

Under the colors tab you can change how the color of the background appears. Figure 1.7 shows the colors tab.

- **Color Schemes:** The colors tab allows you to change how the colors of the Autodesk Inventor screen appear. There are predetermined color schemes for you to choose from. You can also create your own schemes. See the on-line help for information on creating your own color schemes.

- **Background Gradient:** When checked, the background of the Autodesk Inventor screen will change from light to dark from the bottom of the screen. When deselected a solid background will be used.

- **Show Reflections and Textures:** When selected and a reflective material or a material that has a texture is applied to a part a reflection and/or texture will appear on the part.

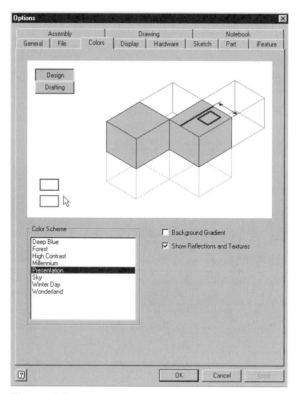

Figure 1.7

DISPLAY TAB

Here you can adjust how the parts will appear. Figure 1.8 shows the available options to change how the parts will look. Depending upon your video card and your requirements, you will want to try different settings to achieve maximum video performance.

- **Wireframe Display Mode:** Sets the preferences for wireframe display.

    Depth Dimming: When checked and in wireframe display mode, the edges furthest away from sight will be more dimmed to better convey the depth of the part.

    Active: Action will be taken on the active part.

    Silhouettes: When checked and in wireframe display mode, the active part will be shown with silhouette edges on.

    Hidden Edges: When checked and in wireframe display mode, the active part's hidden lines will be dimmed.

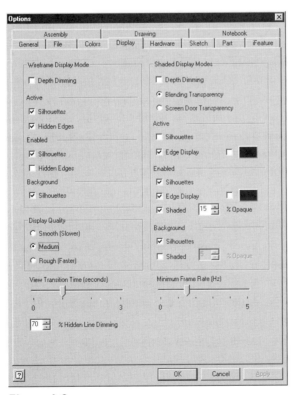

Figure 1.8

Enabled: Action will be taken on the enabled parts.

Silhouettes: When checked and in wireframe display mode, enabled parts of an assembly will be shown with silhouette edges on.

Hidden Edges: When checked and in wireframe display mode, the enabled parts of an assembly will show their hidden lines dimmed.

Background: Action will be taken on the parts that are not enabled.

Silhouettes: When checked and in wireframe display mode, parts that are not enabled will be shown with silhouette edges on.

- **Display Quality:** Select one of three options to set how the parts and assemblies will be displayed when rotated. Generally, the smoother the resolution, the longer it takes to redisplay the model when changes are made. When working with a very large or complex model, you may want to lower the quality of the display to speed up operation. For example, the Rough setting temporarily simplifies detail on large parts but updates faster. The Smooth setting temporarily simplifies fewer details but updates slower.

    Smooth (Slower): The parts will look better but will require more time for the parts to regenerate.

    Medium: An intermediate setting will improve regeneration but will sacrifice some quality during rotation.

    Rough (Faster): When parts are rotated, detail will be lost and regeneration will be done when the rotation stops.

- **Shaded Display Modes**

    Depth Dimming: When checked and in shaded display mode, the faces and edges furthest away from sight will be dimmed to better convey the depth of the part.

    Blending Transparency: Blending specifies a high quality transparency display that is achieved by averaging the colors of overlapping objects.

    Screen Door Transparency: Screen Door specifies a lower quality transparency display that is achieved by using a pattern that allows the color of the hidden object to show through.

    Active: Action will be taken on the active part.

    Silhouettes: When checked and in shaded display mode, the active part will be shown with silhouette edges on.

    Edge Display: When checked and in shaded display mode the active part's edges will be displayed. To display the edges in a specified color, check the box to the right and choose a color from the Color dialog box.

Enabled: Action will be taken on the enabled parts.

Silhouettes: When checked and in shaded display mode, the enabled parts of an assembly will be shown with silhouette edges on.

Edge Display: When checked and in shaded display mode, enabled parts edges will be displayed as the same color of the part. To display the edges in a specified color, check the box to the right and choose a color from the Color dialog box.

Shaded: When checked and in shaded mode, enabled parts of an assembly will be shaded when another part is active. When shaded is not checked, enabled parts will be displayed in wireframe when another part is active.

% Opaque: When shaded is checked, the opacity of the shaded parts can be set by typing in a value or by using the up and down arrow to adjust the value.

Background: Action will be taken on the parts that are not enabled.

Silhouettes: When checked and in shaded display mode, parts that are not enabled will be shown with silhouette edges on.

Shaded: When checked and in shaded mode, parts of an assembly that are not enabled will be shaded when another part is active. When Shaded is not checked, parts that are not enabled will be displayed in wireframe when another part is active.

% Opaque: When Shaded is checked, the opacity of the shaded parts can be set by typing in a value or by using the up and down arrow to adjust the value.

- **View Transition Time (seconds):** Set in seconds, the amount of time that a view takes to transition to a new view.

- **Minimum Frame Rate (Hz):** Specify the frame rate that you want for Inventor to update the display during interactive operations like rotate, zoom, and pan. A lower number will cause Inventor to draw everything in the view; time will be sacrificed. A higher number will improve performance but quality will be sacrificed. When the new display is complete or paused, the part(s) will regenerate back to normal.

- **% Hidden Line Dimming:** Sets the percent of dimming for hidden edges when one or more of the Hidden Edges check boxes is selected. A higher number will cause the hidden lines to appear lighter.

HARDWARE TAB

Here you can adjust how Inventor will interact with your video card. Since Inventor is very dependent on your video card, take time to make sure that you are running a supported video card and running the recommended video drivers. If you are experiencing video related issues, experiment with the options on the Hardware tab as shown in Figure 1.9. For more information about video drivers, pick Graphics Drivers from the Help pull down menu.

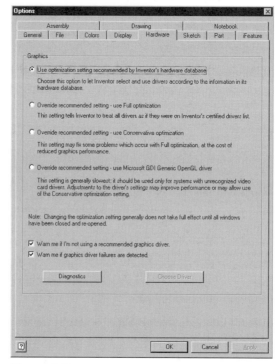

Figure 1.9

DESIGN SUPPORT SYSTEM

Autodesk Inventors help system goes beyond basic command definition by offering assistance while you design. The following help mechanisms makes up the Design Support System: Help Topics, Visual Syllabus, What's New, Tutorials, Design Professor, Design Doctor, Sketch Doctor, and Autodesk Online. To access the help system, use one of these methods.

- Press the F1 key and the help system will give you help with the operation that is active.
- Select a help option from the Help pull down menu.

- Select a help option from the right side of the standard toolbar.
- Select the "?" icon in any dialog box.

The visual syllabus and tutorials guide you through the process to complete chosen tasks. To start a tutorial, pick Tutorials from the Help pull down menu, Figure 1.10 shows the first page for the tutorial on projects. The Design and Sketch Doctor will appear when a problem exists with the operation that you are trying to complete. Another way to get into the Design and Sketch Doctor is to click on the cross on the right side of the Standard toolbar when it appears red. This is alerting you to a problem. Once the Design or Sketch Doctor is open, follow the options that are given to you to resolve the problem.

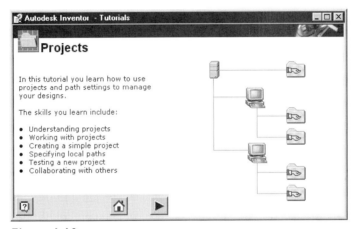

Figure 1.10

USER INTERFACE

Once you start a new Autodesk Inventor file, your screen (application window) will look similar to Figure 1.11. Autodesk Inventor supports multiple documents, meaning that you can open multiple Autodesk Inventor files at the same time in a single Autodesk Inventor session. You can switch between the documents by picking the file name from the Windows pull down menu or the files can be arranged to fit the screen or cascaded. If the files are arranged or cascaded, click on the file to activate it. Only one file can be active at a one time. The screen is divided into the following areas.

- **Pull Down Menus:** Access tools via text menus.
- **Standard Toolbar:** Access basic Windows and Autodesk Inventor tools.
- **Command Bar:** Here Autodesk Inventor tells you the current settings and allows quick editing of objects.

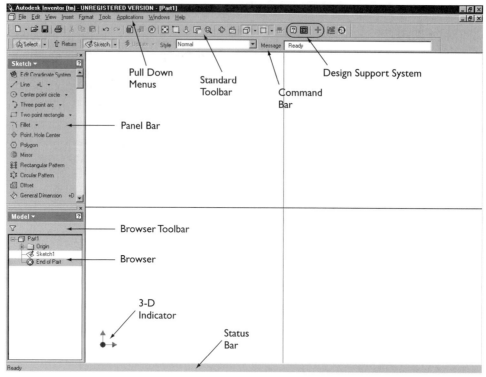

Figure 1.11

- **Panel Bar:** Here is where the majority of Autodesk Inventor tools will be activated. The set of tools located in the panel bar will change to reflect the mode that you are working in.
- **Browser Toolbar:** Allows the user to change how the browser looks. The default settings will be used for this book.
- **Browser:** The browser shows the history of how the file was created, the browser can also be used to edit objects.
- **Graphics Window:** The graphics of the current file will be displayed here.
- **Status Bar:** On the status bar Autodesk Inventor will give you text messages about where you are.

TOOLBARS AND PANEL BAR

The toolbars can be moved to another docked or undocked location. To move a toolbar, pick the two horizontal lines on the top of the toolbar, keep the mouse button depressed and move the mouse to a new location. You will get a preview of what it will look like. Toolbars can be turned on and off by going under the View pull down menu and select Toolbar. The tools in the Panel Bar will adjust to the mode that you are working in (part, assembly, sheet metal etc.). You can manually change the status of the Panel Bar by selecting on the title line of the Panel Bar that has a down arrow, or right click in the Panel Bar and select the one that you want to work with. The tools in the panel bar by default are in non-expert mode; they show a line of text describing the icon. To turn the expert mode on, either select the title area of the menu or right click in the Panel Bar and click on Expert.

COMMAND ENTRY

With Autodesk Inventor there are many ways to issue a command. In the following sections you will learn how to start a command. There is no right or wrong way to start a command; with experience you will develop your own preference for issuing commands.

TOOLBARS AND PANEL BAR

In the last section you learned about toolbars and the Panel Bar that Autodesk Inventor uses. To use a tool from a toolbar move the mouse over the desired tool and a tool tip will appear with the name of the tool. Figure 1.12 shows a portion of the Sketch toolbar. Some of the icons in the Panel Bar have a small down arrow in the lower right corner; select the arrow to see additional tools that are available. Once you are on the icon that you want, release the mouse button and the tool will be activated.

Figure 1.12

SHORTCUT MENUS

Autodesk Inventor also uses shortcut menus. Shortcut menus are text menus that pop up when you right mouse click. The shortcut menus are context sensitive, depending upon where and when you right mouse click you will get a context menu that will list the tools or options that are relevant at that time.

WINDOWS SHORTCUTS

Another method for activating tools is to use Windows shortcut keys. Windows shortcuts use a two key combination. Select the two keys at the same time to activate the operation.

- CTRL C = **Copy**
- CTRL N = **Create a new document**
- CTRL O = **Open a document**
- CTRL P = **Print the active document**
- CTRL S = **Save the current document**
- CTRL V = **Paste**
- CTRL Y = **Redo**
- CTRL Z = **Undo**

HOT KEYS

Autodesk Inventor also has keystrokes that are preprogrammed called hot keys. To start a command via a hot key, press the desired key(s). The following list shows the preprogrammed keys.

- F1 = Help
- F2 = Pan the screen
- F3 = Zoom in or out on the objects on the screen
- F4 = Rotate the objects on the screen
- F5 = Return to the previous view
- SHIFT + F5 = Next View
- B = Add a balloon to a drawing
- C = Add an assembly constraint
- C = Add a center mark when working on a drawing
- D = Add a general dimension to the sketch or drawing
- E = Extrude a sketch
- F = Add a feature control frame to a drawing
- H = Brings up the Hole dialog for creating holes
- L = Draw a line
- O = Add an ordinate dimension
- P = Place a component into the current assembly
- R = Revolve a sketch
- S = Make a plane the active sketch
- T = Add a tweak to a part in the current presentation file
- ESC = Abort command
- DELETE = Delete the selected objects
- BACKSPACE = Will clear the last sketch selection as long as the command is active
- SHIFT + right mouse click = Selection tool menu
- SHIFT + Rotate tools = Auto rotate, puts the contents of the screen into hands free rotation. Press the left mouse button to stop the rotation.
- SPACE BAR = While in the rotate command, will toggle the common view (glass box) on and off
- TAB = Alternate between input fields

- CTRL and SHIFT keys are used to add and remove objects from a selection set
- CTRL + ENTER = Disable inferencing when committing precise input sketch points

UNDO AND REDO

While working you may want to undo the action that you just performed or undo the undo that you just did. The Undo tool backs up Autodesk Inventor one function at a time. If you Undo too far you can use the Redo tool to move forward one step, but only one step. The zoom, rotate, and pan tools do not affect the Undo and Redo tool. To start the tools either use the shortcut keys; CTRL Y for Redo or CTRL Z for Undo, or select the tool from the Standard toolbar as shown in Figure 1.13. The Undo tool is to the left and the Redo tool is to the right.

Figure 1.13

SKETCHING AND PART APPLICATION OPTIONS

Before learning about sketching look at the options in Autodesk Inventor that will affect sketching. While learning Autodesk Inventor, refer back to these option settings to determine the settings that work best for you; there are no right or wrong settings.

SKETCH OPTIONS

Under the Tools pull down menu, select Application Options and the Options dialog box will appear; select the Sketch tab and your screen should look similar to Figure 1.14. Following is a description of the Sketch options. These settings are global and they will effect all open and new Autodesk Inventor documents with the exception of the grid spacing and the 3-D Auto-Bend Radius that are document specific.

- **Constraint Placement Priority**

    Parallel and Perpendicular: When checked and a parallel or perpendicular condition exists while sketching, a parallel or perpendicular constraint will be applied before any other possible constraints that effect the geometry being created.

    Horizontal and Vertical: When checked and a horizontal or vertical condition exists while sketching, a horizontal or vertical constraint will be applied before any other possible constraints that effect the geometry being created.

- **Display**

    Grid Lines: They will toggle on and off both minor and major grid lines on the screen. The grid distance can be set in the current document by clicking on the Tools pull down menu, then select Document Settings, and then from the Sketch tab set the X and Y snap distance.

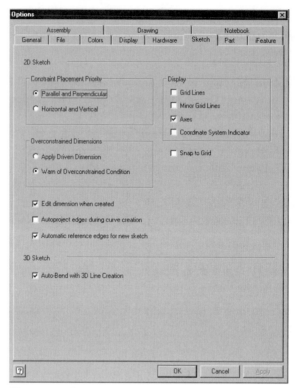

Figure 1.14

> **Minor Grid Lines:** They will toggle on and off the minor grid lines displayed on the screen.
>
> **Axes:** They will toggle on and off the lines that represent the X and Y axis of the current sketch.
>
> **Coordinate System Indicator:** They will toggle on and off the icon that represents the X, Y, and Z axis at the 0,0,0 coordinates of the current sketch.

- **Overconstrained Dimensions**

  > **Apply Driven Dimensions:** When checked and dimensions are added that would over-constrain the sketch, the dimension is added as a driven (reference) dimension.
  >
  > **Warn of Overconstrained Condition:** When checked and dimensions are added that would over-constrain the sketch, the dimension dialog box appears warning of the condition, and then the dimension is not added.

- **Snap to Grid:** When checked, endpoints of sketches will snap to the intersections of the grid as the mouse moves over them.
- **Edit Dimensions When Created:** When checked, the values of dimensions will be edited in the dimension edit dialog box immediately after the dimension is positioned.
- **AutoProject Edges During Curve Creation:** When checked and while sketching objects that are in a different plane, "scrubbing" an object will automatically project it onto the current sketch. Autoproject can also be toggled on and off while sketching by clicking the right mouse button and selecting Autoproject.
- **Automatic Reference Edges for New Sketch:** When checked and you create a new sketch, all of the edges that define that plane will be automatically projected onto the sketch plane as reference geometry.
- **3-D Sketch**

    Auto-Bend with 3-D Line Creation: When checked, a tangent arc will automatically be placed between two 3-D lines as they are sketched. The radius or the arc can be set in the current document by clicking on the Tools pull down menu, then selecting Document Settings, and then setting the Auto-Bend Radius from the Sketch tab.

PART OPTIONS

Under the Tools pull down menu select Application Options, and the Options dialog box will appear. Select the Part tab and your screen should look similar to Figure 1.15. Following is a description of the Part options. These setting are global; they will affect all open and new Autodesk Inventor documents.

- **Sketch on New Part Creation**

    No New Sketch: When checked, no sketch plane is set when a new part is created.

    Sketch on X-Y Plane: When checked, the X-Y plane is set as the current sketch plane when a new part is created.

    Sketch on Y-Z Plane: When checked, the Y-Z plane is set as the current sketch plane when a new part is created.

    Sketch on X-Z Plane: When checked, the X-Z plane is set as the current sketch plane when a new part is created.

- **Parallel View on Sketch Creation:** When checked, the view orientation will automatically be changed to look directly at the sketch plane.
- **Auto-Hide In-Line Work Features:** When checked, work features that are consumed by another work feature will automatically be hidden.

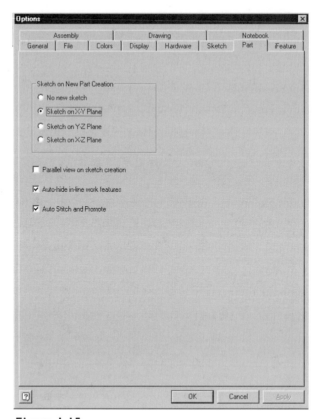

Figure 1.15

- **Auto Stitch and Promote:** When checked, surface models that are imported into Inventor will be stitched together and automatically converted to an Inventor part.

UNITS

Autodesk Inventor uses a default unit of measurement for every part, assembly, and drawing file. The default unit is set from the template file that it was created from. When specifying numbers in dialog boxes and no unit is given, the default unit will be used. The default units for an active part or assembly document can be changed by clicking on the Tools pull down menu then select Document Settings and then the Units tab. A Document Settings dialog box will appear as shown in Figure 1.16. When the unit system values are changed all the existing values in that file will also be changed. The dimension style and active drafting standard set units in a drawing.

Drawing settings will be covered in chapter 4. The default units can be overridden on any number by typing in the desired unit. For example, if you were working on an "inch" file and you placed a horizontal dimension whose default value was "2 in" you could type in "50 mm." When the dimension is displayed on the screen it will appear in the default units. For the previous example "2 in" would be displayed on the screen. When the dimension would be edited the overridden unit would be displayed.

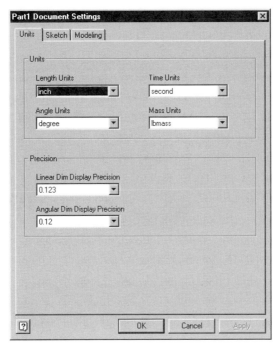

Figure 1.16

TEMPLATES

After making changes to how you would like your parts and assemblies to appear, you can create your own templates. Then when you create a new part or assembly, start the new file based on that template. You may add templates to your existing Autodesk\Inventor "version number"\Templates folder or create a new subdirectory under the templates folder and add them there. After you create the new template subdirectory, a new tab will appear when creating new files once you have placed a file in the folder.

CREATING A PART

The first step in creating a part is to start a new part file or create a new part file in an assembly. The first three chapters of this book will deal with creating parts in a new part file, and chapter 5 will cover assemblies. To create a new part file select the New icon from the "What To Do" section and then select the Standard.ipt icon from the Default template tab as shown in Figure 1.17. There are also two other methods for creating a new part. You can select New from the File pull down menu or select the down arrow of the New icon and select Part from the left side of the Standard toolbar. After issuing the new part tool, Inventor's screen will then change to reflect the new part environment.

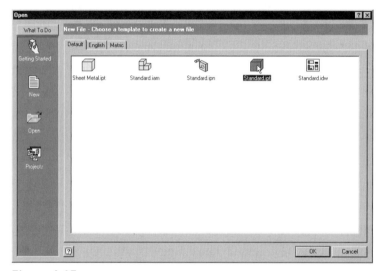

Figure 1.17

SKETCHES AND DEFAULT PLANES

Before you can start sketching, you must have an active sketch. A sketch is a plane that 2-D objects are sketched on. Any planar face or workplane on a part can be made the active sketch. A sketch is automatically set by default when you create a new part file. The default plane that the sketch is created on can be changed by selecting the Tools pull down menu then select Application Options; from the Part tab, pick the sketch plane that new parts should be defaulted to. Also each time a new Inventor part is created, there are three planes (XY, YZ, and XZ), three axes (X, Y, and Z), and the center point that are automatically created. At the middle of the workplanes is a work axis and at the intersection of the three planes is the center (origin) point. By default visibility is turned off to these planes, axes, and center point. To see the planes, axes, or center point, expand the Origin entry in the browser by clicking on the + to the left side of the text. Then move the mouse over the names and they will appear in the graphics area. Figure 1.18 shows these planes, axes, and center point in the graphics area shown in an isometric view and the Browser is shown with the Origin ex-

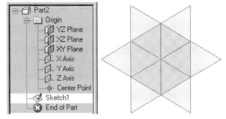

Figure 1.18

panded. To leave the visibility of the planes or axes on, in the browser right click the mouse while over the name and select Visibility. These default planes can also be made the active sketch.

NEW SKETCH

To make a planar face of a part, a workplane, or an existing sketch on the part the active sketch, you need to issue the 2-D Sketch tool. The New Sketch tool can be started by doing one of the following:

Figure 1.19

- Pick the Sketch tool from the Command Bar as shown in Figure 1.19. Then select a face of a part, a workplane, or an existing sketch from the Browser. You may also do it the opposite way by selecting a face of a part, a workplane, or an existing sketch from the Browser and then picking the Sketch tool from the Command Bar.

- Press the hot key S and then select a face of a part, a workplane, or an existing sketch from the Browser or select a face of a part, a workplane or an existing sketch from the Browser and then Press the hot key S.

- While not in the middle of an operation, right click in the graphics window and select New Sketch from the context menu as shown in Figure 1.20 or select a face of a part, a workplane, or an existing sketch from the Browser and then right click in the graphics window and select New Sketch from the context menu.

Figure 1.20

After the active sketch has been created the X and Y axes will be automatically aligned on this plane and you can then begin to sketch.

 Note: This book assumes that when you installed Inventor you selected inch as the default unit. If mm was selected as the default unit, then select the Standard.ipt (in) template file from the English tab. This book will use the XY plane as the default sketch plane.

SKETCH THE OUTLINE OF THE PART: STEP 1

As stated at the beginning of the chapter, 3-D parts usually start with a 2-D sketch of the outline shape of the part. A sketch can be created with lines, arcs, circles, splines, or any combination. This section will cover sketching strategies, tools, and techniques.

SKETCHING OVERVIEW

When deciding what outline to start with, analyze what the finished shape will look like. Look for the shape that best describes the part. When looking for this outline, try to look for a flat face. It is usually easier to work on a flat face than on a curved edge, which can be difficult for new users. However, as you gain modeling experience, reflect back on how the model was created and think about other ways that the model could have been built. There is usually more than one way to generate a given part.

When sketching, draw the geometry so it looks *close* to the desired shape and size; you do not need to be concerned about exact dimensional values. Inventor allows islands in the sketch (closed objects that lie within another closed object). An example would be a circle that is drawn inside of a rectangle. When the sketch would be extruded the island would become a void in the solid. A sketch can consist of multiple closed objects that are coincident.

Here are some guidelines that will help you generate good sketches.

Tip for Sketching: Select an outline that represents the part best. It is usually easier to work from a flat face.

Draw the geometry close to the finished size. For example, if you want a 2″ square, do not draw a 200″ square.

Create the sketch proportionately in size to the finished shape. When drawing the first line segment use the precise input dialog to sketch the first line to size. This will give you a guide on how to sketch the rest.

When sketching, notice that the distance and angle of the object being sketched will appear at the bottom of the screen. Use this information as a guide.

Draw the sketch so that it does not overlap. The geometry should start and end at the same point.

Do not allow the sketch to have a gap; all the connecting endpoints should be coincident.

Keep the sketches simple. Leave out fillets and chamfers. They can easily be placed as features after the sketch is turned into a solid. The simpler the sketch, the fewer number of constraints and dimensions will be required to constrain the model.

If you want to create a solid, the sketch must form a closed shape; if it is open, it can be turned into a surface only.

SKETCHING TOOLS

Before we start sketching the part, let's look at the sketching tools that are available. By default the panel bar will show the Sketch tools. Figure 1.21 shows the Sketch Toolbar with the expert mode turned off (text descriptions shown). As you become more proficient with Inventor, you can turn off the text by either selecting on the title area of the menu or right click in the Panel Bar and click on Expert. Following is a description of the sketch tool.

- **Edit Coordinate System:** This realigns the sketch coordinate system to existing objects. Arrows on the screen will indicate the directions of the X axis and the Y axis.
- **Line:** This command will draw line segments; an arc segment can be drawn from the endpoint of a line. The next section will cover the actual sketching techniques.

    Spline: Draw a spline.
- **Center Point Circle:** Draw a circle by selecting a center point for the circle and then a point on the circumference of the circle.

    Tangent Circle: Draw a circle that will be tangent to three lines or edges by selecting the lines or edges.

    Ellipse: Draw an ellipse by selecting a center point for the ellipse, a point on the first axis, and another point that will lie on the ellipse.
- **Three Point Arc:** Draw an arc by selecting its end points, then a point that will lie on the arc.

    Tangent Arc: Draw an arc that is tangent to an existing line or arc by selecting the endpoint of the line or arc then select the other endpoint of the arc.

    Center Point Arc: Draw an arc by picking a center point for the arc then pick a start and end point.
- **Two Point Rectangle:** Draw a rectangle by picking a point and then pick a point to define the opposite side of the rectangle. The edges of the rectangle will be horizontal and vertical.

    Three Point Rectangle: Draw a rectangle by picking two points that will define an edge and then select a point to define the third corner.
- **Fillet:** Create a fillet between two lines, two arcs, or a line and arc at a specified radius between to nonparallel lines. If two parallel lines are selected, the fillet will be created between them without specifying a radius.

    Chamfer: Create a chamfer between lines. There are three options to create a chamfer: both sides are equal distance, two defined distances, or a distance and angle.

- **Point, Hole Center:** Creates a Hole Center that will be used to place a hole or Points that can be used as vertices for other objects, that is, splines.
- **Polygon:** Create an inscribed or a circumscribed polygon with the number of faces that you need.
- **Mirror:** Mirror the selected objects about a centerline. A symmetry constraint will be applied to the mirrored objects.
- **Rectangular Pattern:** Creates a rectangular array of a sketch with a user defined number of rows and columns.
- **Circular Pattern:** Creates a circular array of a sketch with a user defined number of copies and spacing.
- **Offset:** Create a duplicate of the select objects that are a given distance away. By default an equal distance constraint is applied to the offset objects.
- **General Dimension:** Create a parametric or a driven dimension.
- **Auto Dimension:** This command will automatically place dimensions and constraints to a selected sketch. Auto dimensioning will be covered in chapter 6.

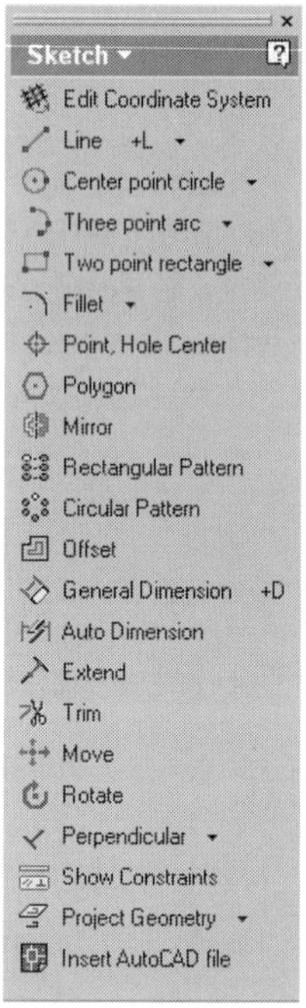

Figure 1.21

- **Extend:** Extend the selected object to the next object it finds. Select near the end of the object that you want extended. While in the Extend tool hold down the SHIFT key to trim objects.
- **Trim:** Trim the selected object to the next object it finds. Select near the end of the object that you want trimmed. While in the Trim tool hold down the SHIFT key to extend objects.
- **Move:** Moves the selected sketch from one point to another point. A copy of the sketch can be made by selecting Copy in the Move dialog box. If the moved sketch has objects that are constrained to them, they will also be moved.
- **Rotate:** Rotates the selected sketch about a specified point. A copy of the sketch can be made by selecting Copy in the Rotate dialog box. If the rotated sketch has objects that are constrained to them, they will also be rotated.
- **Constraints:** These tools (accessed from the flyout) will place geometric constraints on the geometry. Constraints will be covered in detail later in this chapter.

- **Show Constraints:** After issuing the Show Constraints tool, select the object(s) whose constraints you want to see. When the constraints are visible, they can be deleted by right clicking on a constraint and then selecting Delete from the context context menu. Showing constraints will be covered in detail later in this chapter.

- **Project Geometry:** Selected edges, vertices, and work features will be projected onto the current sketch. The projected edges, vertices, and work features can be used as part of the sketch or used to dimension to. The projected edges, vertices and work features are constrained to the original edge, vertex, or work feature. If the
parent object changes or moves so will the projected object.

    Project Cut Edges: Edges that lie on the section plane will be projected onto the sketch of the new part. Geometry is only projected if the uncut part would intersect the sketch.

    Project Flat Pattern: When working on a sheet metal part, the sections of the flat pattern can be projected onto the part. Sketches can then be dimensioned to both the part and projected flat.

- **Insert AutoCAD File:** Insert an AutoCAD 2-D file onto the current sketch.

USING THE SKETCH TOOLS

After starting a new part, use the sketch tools to draw the shape of the part. As you are sketching Inventor gives you visual feedback to what is happening on the screen and in the Message area of the Command Bar. After issuing a sketch tool, follow the prompts in the message area. To start sketching, issue the tool that you want, and then in the graphics area select a point that you want with the left mouse button and follow the prompt in the Message area on the command line. Each section below will introduce a technique that can be used to create a sketch.

Line Tool

The line tool is one of the most powerful tools that you will use to sketch. Not only can you draw lines but you can also draw an arc from the endpoint of a line segment. After issuing the Line tool you will be prompted to select a first point, select a point in the graphics area and then pick a second point. You can continue drawing line segments or from the endpoint of a line segment or arc move the mouse over the end point and a small circle will appear at that endpoint. Figure 1.22 shows how the endpoint of a line segment looks when the mouse is moved over it. Click on the small circle and then with the left mouse button pressed down, move the mouse in the direction that you want the arc to go. Depending upon how you move the mouse, up to eight different arcs could be drawn. The arc will be tangent to the horizontal or vertical edges that are displayed from the selected endpoint. Figure 1.23 shows an arc being drawn that is normal to the sketched line.

Tip: When sketching, check the Status Bar (bottom of the screen) to see the length and angle of the objects that you are drawing.

Figure 1.22   **Figure 1.23**

Inferred Points

As you are sketching, dashed lines will appear on the screen. These dashed lines represent the endpoints of lines and arcs that represent their horizontal, vertical, or perpendicular positions. As the mouse gets close to these inferred points, it will snap to that location. If that is the point that you want, select that point—otherwise continue to move the mouse until it reaches the desired location. When inferred points are selected *no* constraints (geometric rules: horizontal, vertical, collinear, etc.) are applied from these inferred points. Using inferred points helps create more accurate sketches. Figure 1.24 shows the inferred points from two endpoints that represent the horizontal and vertical points of those endpoints.

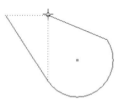

Figure 1.24

Automatic Constraints

As you are sketching small glyphs (small symbols) will appear that represent constraint(s) that will be applied to the object. If you do not want a constraint to be applied, hold down the CTRL key when the point is selected. Figure 1.25 shows a line being drawn from the arc and the line will be tangent to the arc and parallel to the angled line. The glyphs will be shown near the object that the constraint is coming from. Constraints will be covered in detail in the next section.

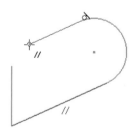

Figure 1.25

Scrubbing

When sketching, you may prefer to have a different constraint applied than the one that automatically appears on the screen. For example, you may want that line to be perpendicular to a given line instead of being parallel. The technique to accomplish this task is called *scrubbing*. When sketching you can apply a constraint from another object by moving the mouse over the other object that the constraint should be applied to and then move the mouse back to its location to pick its next point. The glyph should change to show the constraint that would be applicable to the line that was scrubbed. The line color of the scrubbed line will become darker, representing that it is the object that the constraint is being applied to. Figure 1.26 shows the top horizontal line being drawn with a perpendicular constraint that was scrubbed from the left vertical line. Without scrubbing the left vertical line, the applied constraint would have been parallel to the line along the bottom line.

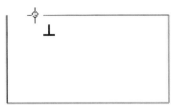

Figure 1.26

Precise Input

When drawing line segments, it may be easier to draw them at a specified length or angle. These values can be entered into the Precise Input dialog box. To open the Precise Input dialog box, select the View pull down menu, then select Toolbars, and then select Precise Input or right click on either the Standard or Command bar and pick Precise Input from the list. Before entering values into the dialog box, you need to start a sketch tool-like line. Then select the method of input you want and then type the values into the Precise Input dialog box. Figure 1.27 shows the Precise Input dialog box with the input options displayed.

Figure 1.27

Selecting Objects

After sketching objects you may need to move, rotate, or delete some or all of them. There are two methods to select objects for editing.

1. Select them individually and to add objects to a selection, set by holding down the CTRL or SHIFT key while selecting the objects. Selected objects can be removed from a selection set by holding down the CTRL or SHIFT key and selecting them. As objects are selected their color will change to represent that they have been selected.

2. While not in a command, draw a box around the objects. If the box is drawn from left to right, only the objects that are fully enclosed in the box will be selected. If the box is drawn from right to left, all the objects that are fully enclosed in the box and the objects that are touched by the box will be selected.

3. You may use a combination of steps 1 and 2 to select the objects that you want to work with.

Deleting Objects

To delete objects, first select them and then either press the DELETE key or right mouse click and select Delete from the context menu.

TUTORIAL 1.2—SKETCHING

1. Start a new part file based on the default Standard.ipt file.
2. Sketch the geometry as shown in Figure 1.28; the dimensions are shown for reference only. The two angle lines should be parallel; the top angle line should be tangent to the top arc, and the left quadrant of the arc should stop at the center point of the arc.
3. Save the file with the following name:
   \Inventor Book\TU1-2.ipt. When asked to exit the sketch mode, select OK.
4. Close the file.
5. Create a new part file and practice sketching different shapes. Use the Precise Input, trim, extend, different arc creation methods. Use the different selection methods and move, copy, and delete some of them.
6. When you are done practicing using the sketching tools, close the file.

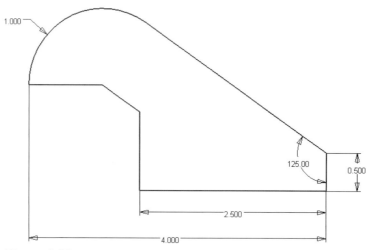

Figure 1.28

CONSTRAINING THE SKETCH: STEP 2

After the sketch is drawn you will want to add geometric constraints to the sketch. Constraints are used to apply behavior to specific objects or between objects. As each constraint or dimension is placed, the sketch is being constrained. A fully constrained sketch is a sketch whose objects cannot move. An example of a constraint would be to apply a vertical constraint to a line so that it will always be vertical. You could apply a parallel constraint between two lines to make them parallel to one another. As one of the line's angles would change so would the other. A line and arc could have a tangent constraint applied to them. As stated in the last section, constraints can be applied while the objects are being sketched. You may also need to apply additional constraints to control the sketch. By default, after being sketched, objects are free to move in their sketch plane. To prevent a sketch from moving in the plane, a point, points, or an edge or edges must be fixed using the Fix constraint. After a point or edge is fixed, the objects will then turn a darker color after they are constrained (cannot move). A sketch cannot be fully constrained until at least one Fix constraint is placed.

DRAGGING A SKETCH

To help determine if an object is constrained you can try to drag it to a new location. While not in a command, select a point or an edge and with the left mouse button depressed, drag it to a new location. If the geometry stretched is under-constrained, if the entire sketch moves a Fix constraint needs to be added. If dimensions are on the object, they too will prevent the object from stretching. Dragging constraints take precedence over dimensions. For example if you drew a square that had two horizontal and vertical constraints applied to it, then select a point on one of the corners. You could change the size of the square but the lines would maintain their horizontal and vertical constraints.

CONSTRAINT TYPES

Figure 1.29

To apply a constraint, select from the constraint in the Sketch toolbar. Each constraint is described in the next section. Figure 1.29 shows the constraint types and the symbol that represents them.

Perpendicular Constraint: Lines will be positioned 90° to one another; that is, the first line you sketched will stay and the second will rotate until the angle between them is 90°.

Parallel Constraint: Lines will be positioned so they are parallel to one another; that is, the first line you sketched will stay and the second will move to become parallel to the first.

Tangent Constraint: An arc and a line or two arcs will become tangent.

Coincident Constraint: A gap between two endpoints of arcs and or lines will become closed.

Concentric Constraint: Arcs and or circles will share the same center point.

Collinear Constraint: Two selected lines will line up along a single line; if the first line moves, so will the second. The two lines do not have to be touching.

Horizontal Constraint: Lines are positioned parallel to the X axis, or a Horizontal constraint can be applied between the center points of arcs or circles. The center points will then share the same horizontal axis.

Vertical Constraint: Lines are positioned parallel to the Y axis, or a Vertical constraint can be applied between the center points of arcs or circles. The center points will then share the same vertical axis.

Equal Constraint: If two arcs or circles are selected, they will then have the same radius or diameter. If two lines are selected, then the line segments will become the same length. If one of the objects changes so will the other object that the equal constraint has been applied to. If the equal constraint is applied after one of the arcs, circles, or lines have been dimensioned, the second arc, circle, or line will take on the size of the first one.

Fix Constraint: Applying a fixed point(s) will prevent the endpoints or edges of objects from moving. The fixed point overrides any other constraint. Any endpoint or segment of a line, arc, circle, spline segment, or ellipse can be fixed. If you select near the endpoint of an object, the endpoint will be locked from moving. If you select near the midpoint of a segment the entire segment will be locked from moving. Multiple points in a sketch can be fixed. If applying constraints and the profile is moving in directions that are undesirable, you can apply fix constraints to hold the endpoints of the objects in place. You can remove a fix constraint as needed. Deleting constraints will be covered later in this chapter.

Symmetry: Selected geometry will be symmetric about another, line, centerline or edge. Symmetry will be covered in chapter 6.

 Note: If a point or edge is not fixed in the sketch, then the sketch cannot become fully constrained.

SHOWING AND DELETING CONSTRAINTS

To see the constraints that are applied to an object, use the Show Constraints tool from the Sketch Panel Bar; see Figure 1.30. After issuing the Show Constraints tool, select an object and a row of constraint icons will appear. As you move the mouse over the constraint icon, the objects that are linked to that constraint will highlight. Figure 1.31 shows the mouse over the parallel constraint; the larger angled line is highlighted to show that it is the line that is parallel to the smaller angled line. To close the row of constraint icons select on the "x" on the right side of the row. To show all the constraints for all the objects on the current sketch, right click in the graphics area and from the pop-up menu, select Show All Constraints. To hide all the constraints right click in the graphics area and from the pop-up menu, select Hide All Constraints. To delete a constraint, first show it using one of the methods just described. Then select on the constraint icon, then right

Figure 1.30

mouse click in the graphics area and from the context menu select Delete. Figure 1.32 shows the parallel constraint being deleted.

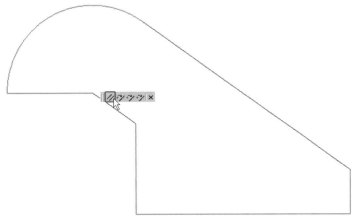

Figure 1.31

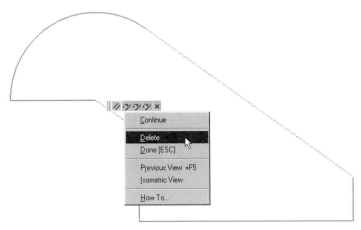

Figure 1.32

ADDING CONSTRAINTS

If you need to add a constraint, select the particular constraint from the Sketch toolbar or right-click in the graphics area and from the context menu as shown back in Figure 1.29. Pick the select Create Constraint and select the constraint you want to apply from the context menu. Then select the object or objects to apply the constraints. As stated at the beginning of this section, constraints set behavior or relationships to the sketch. As the constraints are added the number of dimensions or constraints that are required to fully constrain the sketch will be decreased. First apply a fix constraint and then as objects become fully constrained their color will change. Autodesk Inventor will not allow you to over-constrain the sketch or add duplicate constraints. If you add a constraint that would conflict with another, you will be warned that this constraint would over-constrain the sketch. An example would be to add a vertical constraint to a line that already has a horizontal constraint. If you try to add a constraint to an object that already exists, you will be alerted at the command line that this constraint has already been applied.

ADDING DIMENSIONS: STEP 3

The last step to constraining a sketch is to add dimensions to it. The dimensions you place will control the size of the sketch and will also appear in the drawing views when they are generated. Try to avoid having extension lines that go through the sketch, as this will require more clean up when drawing views are generated. When selecting lines to place dimensions, click near the side where you anticipate the dimensions originating from in the drawing views. All parametric dimensions are created with either the General Dimension or Automatic Dimension command. General dimensioning will be covered in the next section and automatic dimensioning will be covered in chapter 6.

GENERAL DIMENSIONING

The General Dimensioning tool can create a linear, angle, radial, or diameter dimension one at a time. To start the General Dimension tool, issue it from the Sketch Toolbar as shown in Figure 1.33, or right mouse click in the graphics area and select Create Dimension from the context menu or press the hot key D. Once in the tool you are now ready to place a dimension. When placing dimensions the extension line of the dimension will automatically snap to the nearest endpoint; when an arc or circle is selected, it will snap to its center point. To dimension to a quadrant of an arc or circle, see the procedure on the next page. Before a dimension is created a preview image will appear on the screen showing what type it is and where it will be placed. If the dimension type is not what you want, right mouse click and select the correct style from the context menu. Figure 1.34 shows the context menu that is available when a vertical dimension is previewed but will be changed to an Aligned dimension. Change the dimension type to the correct style. The dimension preview will change to this style, then select a point on the screen to place the dimension. The next four sections will cover how to dimension specific objects and types of dimensioning with the General Dimension tool.

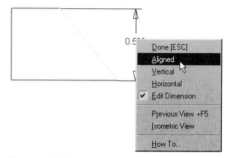

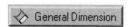

Figure 1.33   Figure 1.34

DIMENSIONING LINES

There are two techniques for dimensioning a line: You can select near two endpoints and then select a location for the dimension. Or if you want to dimension a single line, select the line anywhere and then select a location for that dimension.

DIMENSIONING AN ANGLE

To create an angular dimension, select near the midpoint of the two lines that you want the angle dimension to go between and then select a point for the dimension location.

DIMENSIONING ARCS AND CIRCLES

To dimension an arc or a circle, select on its circumference and then select a location outside of the object. By default, when an arc is dimensioned, the result is a radius dimension; when a circle is dimensioned, the default is a diameter dimension. To change the radial dimension to a diameter or a diameter to radial, press the right mouse button before the dimension is placed and select the other style to place the dimension.

DIAMETRIC DIMENSIONS

Diametric dimensions are used to create diameter dimensions for sketches that represent a quarter outline of the part to be revolved. To create a diametric dimension, draw a sketch that represents a quarter section of the finished part, then draw a centerline that the sketch will be revolved around. Issue the General Dimension tool and pick the centerline and the other point to be dimensioned. Then place the diameter dimension. Figure 1.35 shows a sketch and a centerline with a diametric dimension. Centerlines will be covered in chapter 2 with the Revolve command.

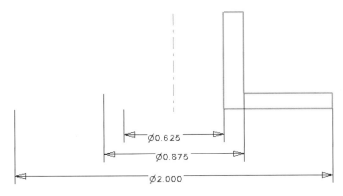

Figure 1.35

DIMENSIONING TO A QUADRANT

To dimension to a quadrant of an arc or circle, select a line that is parallel to the quadrant then move the mouse over the quadrant that should be dimensioned. Keep moving the mouse around the quadrant until the glyph changes to reflect a dimension to the quadrant. Once the quadrant glyph is present select that point and then click to place the dimension.

ENTERING AND EDITING DIMENSION VALUES

After placing the dimension, the Edit Dimension dialog box will appear as shown in Figure 1.36. If the Edit Dimension dialog box does not automatically appear after placing it, pick Application Options from the Tools pull down menu, then from the Sketch tab check "Edit dimension when created." If this option is not checked the dimension will be placed with the default value. You can also right click on the graphics window and select Edit Dimension from the context menu to toggle this option on or off. To edit a dimension that has already been created double click on the value of the dimension and the Edit Dimension dialog box will appear as shown in Figure 1.36. Type in the new value and unit for the dimension; then either

Figure 1.36

press ENTER or select the check mark in the Edit Dimension dialog box. If no unit is entered, the units that the file was created with will be used. When inputting values, type in the exact value; do *not* round up or down. The accuracy shown in the dimension is from the current dimension style. Autodesk Inventor parts are accurate to six decimal places; for example, 1.0625 is more accurate than 1.06. When placing dimensions, it is recommended that you place the smallest dimensions first; this will prevent the geometry from flipping in the wrong direction.

OVER-CONSTRAINED SKETCHES

As stated in the Adding Constraints section, Autodesk Inventor will not allow you to over-constrain the sketch or add duplicate constraints. The same is true when adding dimensions. If you are adding a dimension that would conflict with another constraint or dimension, you will be warned that this dimension would over-constrain the sketch or a driven dimension can be created. A driven dimension is a reference dimension. It is not a parametric dimension; it just reflects the size of the points that it is dimensioned to. A driven dimension will appear with parenthesis around the dimensions value. When you are warned that a dimension would over-constrain the sketch, you can either cancel the command and no dimension will be placed or accept the warning and a driven dimension will be created. To create a driven dimension without being warned, go under the Tools pull down menu, then select Application Options then from the Sketch tab check "Apply Driven Dimension."

TUTORIAL 1.3—CONSTRAINING AND DIMENSIONING

1. Start a new part file based on the default Standard.ipt file.
2. Sketch the geometry as shown in Figure 1.37 and then show all the constraints.

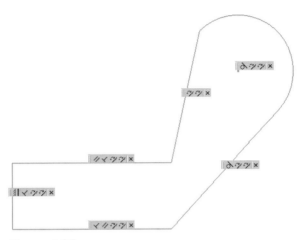

Figure 1.37

3. While not in a command, drag the lower left corner, then drag the outside angle line and then drag the arc. When done, drag the sketch so it again looks similar to Figure 1.37. You may have more or less constraints on your sketch depending upon how you drew it.
4. Add a fix constraint to the lower left corner of the sketch. If the fix constraint icon did not appear on the list of constraints for that corner then reshow all the constraints.
5. If the following constraints exist, delete the vertical and perpendicular constraint on the left vertical line. When done your sketch should look similar to Figure 1.38.
6. While not in a command, drag the top left corner of the left vertical line out so it forms about a 45° angle from the bottom horizontal line.
7. If it does not exist, add a horizontal constraint to the bottom line.
8. Add a tangency constraint between the arc and two lines that touch it.
9. Add a parallel constraint between the two angle lines that come off the arc.
10. If needed, drag the sketch until it looks similar to Figure 1.39.
11. Hide all the constraints.
12. Add dimensions until your sketch looks similar to Figure 1.40.

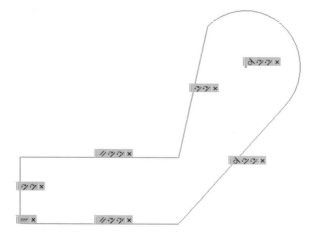

Figure 1.38

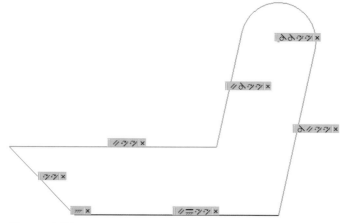

Figure 1.39

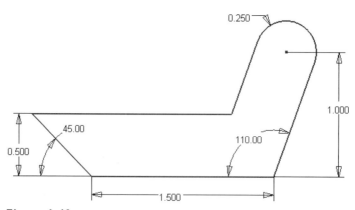

Figure 1.40

13. Delete the 1.000 dimension on the right side of the sketch.
14. Then add a 1.000 dimension from the bottom horizontal line to the top quadrant of the arc.
15. Edit the 0.500 left vertical dimension to 0.375, the 110.00 angle dimension to 125.00 and change the 0.250 radius dimension of the arc to 0.1875. When you are done, your sketch should resemble Figure 1.41.

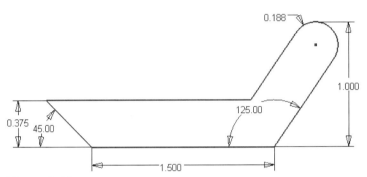

Figure 1.41

16. Save the file with the following name: \Inventor Book\TU1-3.ipt When asked to exit the sketch mode, select OK.
17. Close the file.

PRACTICE EXERCISES

The following exercises are intended to challenge you by providing problems that are open-ended. As with the kinds of the problems you'll encounter in work situations, there are multiple ways to arrive at the intended solution. In the end, your solution should match the sketch that is shown. For each of the following exercises, start a new part file. When you have finished the sketch, save the file using the name supplied in the exercise. You will use the finished sketches in later chapters.

Exercise 1.1—Bracket

Save the finished part (Figure 1.42) as \Inventor Book\Bracket.ipt

 Tip: To align the center of the small outside arcs and the top and bottom horizontal lines, apply a horizontal constraint to the center point of the two smaller arcs on the outside of the sketch to the respective bottom and top horizontal lines. To align the center of the left side of the slot, apply a concentric constraint to the larger arc on the left side of the sketch to the left arc of the slot.

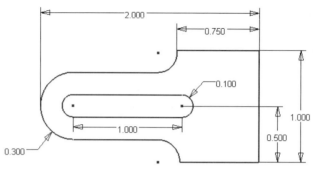

Figure 1.42

Exercise 1.2—Connecting Rod

Save the finished part (Figure 1.43) as \Inventor Book\Connecting Rod.ipt

 Tip: Draw two circles and two lines then trim them to the final shape. To align the center point of the arcs, apply a vertical constraint between the center points of the arcs. To evenly align the top and bottom points of the lines, apply a horizontal constraint between the two top and bottom points of the lines.

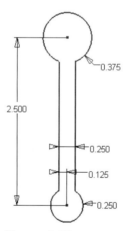

Figure 1.43

REVIEW QUESTIONS

1. Explain what a project file is used for.
2. Autodesk Inventor stores the part, assembly information, and related drawing views in the same file.   True or False?
3. When sketching, by default constraints are *not* applied to the sketch.   True or False?
4. When sketching and a point is inferred, a constraint is applied to represent that relationship.   True or False?
5. Explain how to draw an arc while still in the Line command.
6. A sketch does not need to be fully constrained.   True or False?
7. There can be multiple fix constraints on a sketch.   True or False?
8. Can a constraint be removed? If so, how?
9. When working on an "inch" part, you cannot use metric units.   True or False?
10. When an aligned dimension is required and a vertical dimension appears, can the vertical dimension be changed to an aligned dimension? If so, how?
11. When an arc is dimensioned, is the default a radius or a diameter dimension?
12. Explain how to create a dimension to a quadrant of an arc or circle.
13. To how many decimal places is an Autodesk Inventor part accurate?
14. After a sketch is fully constrained, a dimension's value cannot be changed.   True or False?
15. A driven dimension is another name for a parametric dimension.   True or False?

CHAPTER 2

Viewing, Extruding, Revolving, Editing Features, Sketched Features, and Boolean Operations

After you have drawn, constrained, and dimensioned the sketch, your next step will be to turn that sketch into a 3-D part. This chapter takes you through the options for viewing a part from different viewpoints, as well as creating features using the extrude and revolve tools and editing features.

CHAPTER OBJECTIVES

After completing this chapter, you will be able to:

- view a part(s) from different viewpoints.
- understand what a feature is.
- use the Inventor Browser to edit parts.
- extrude a sketch into a part.
- revolve a sketch into a part.
- edit features of a part.
- edit the sketch of a feature.
- make planes the active sketch.
- create sketched features using one of the three operations: Cut, Join and Intersect.

VIEWING A MODEL FROM DIFFERENT VIEWPOINTS

Up to now you have worked with 2-D sketches in the XY plane (plan view). The next step is to give the 2-D sketch depth in the Z plane. When working in 3-D it is helpful to be able to view the part from different viewpoints as well as zooming in and out and panning the parts on the screen. The next section will guide you through the most common methods of viewing objects from different perspectives and viewpoints. For each of the tools, the physical part(s) is not moving. Your perspective or viewpoint to the part(s) is what creates the movement of the part. If you were in an operation while a viewing operation is issued, that operation will be resumed after the new view is transitioned to.

ISOMETRIC VIEW

While working you can change to an isometric viewpoint by right clicking in the graphics area and then selecting Isometric View from the context menu as shown in Figure 2.1. The view on the screen will transition to a predetermined isometric view. The isometric view that is transitioned to can be redefined while in the Common View (Glass Box) option, Common View will be covered later in this section.

Figure 2.1

CAMERA VIEWS

With Inventor there are two camera viewpoints that can be set: orthographic or perspective. By default the orthographic camera (lines are projected perpendicular to the plane of projection) is set. The perspective camera displays the geometry on the screen converging to point more like how the human eye sees. To change the camera, select either Orthographic Camera from the Standard toolbar as shown in Figure 2.2 or the Perspective Camera that is the option just below Orthographic Camera. For clarity this book will show all the images in the Orthographic Camera. For more options setting the Perspective Camera, use the Design Support System.

Figure 2.2

VIEW TOOLS

To help zoom, pan, and rotate the geometry on the screen, viewing tools are available on the Standard toolbar that are shown in Figure 2.3. Following is a description of the viewing tools going from left to right.

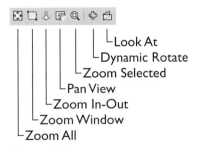

Figure 2.3

Zoom All

To maximize the screen with all the parts that are in the current file, issue the Zoom All tool. The screen will transition to the new view.

Zoom Window

To zoom in on an area that is designated by two points, issue the Zoom Window tool and select the first point and with the mouse button depressed, move the mouse to the second point. A rectangle will appear representing the window. When the correct window is displayed on the screen, release the mouse button and the view will transition to it.

Zoom In-Out

To zoom in or out from the parts, issue the Zoom In-Out tool or press the F3 key. Then in the graphics window press the left mouse key and with the button depressed move the mouse toward you to make the parts appear larger and move it away from you to have the parts appear smaller. If you have a mouse with a wheel you can also zoom in and out on the parts. Roll the wheel toward you and the parts will appear larger; by rolling the wheel away from you the parts will appear smaller.

Pan View

To move the view to a new location, select the Pan View tool or select the F2 key. After issuing the tool, press and hold the left mouse button and the screen will move in the same direction that the mouse moves. If you have a mouse with a wheel, you can hold down the wheel, and while the wheel is depressed, the screen will move in the same direction that the mouse moves.

Zoom Selected

To fill the screen with the maximum size of a selected face or faces, either select the face or faces then issue the Zoom Selected tool or launch the Zoom Selected tool and select the face or faces to zoom to.

Dynamic Rotate

To rotate a part(s) dynamically, issue the Dynamic Rotate model tool. A circular image with lines at the quadrants and center will appear as shown in Figure 2.4. To rotate the parts freely, select a point inside the circle and with the mouse button being pressed move the mouse and the model rotates in the mouse's direction. When you release the mouse button, the model stops rotating. To accept the view orientation, either press the ESC key or right click and select Done from the context menu. Select on the outside of the circle to rotate the model about the center of the circle. To rotate the parts about the vertical axis, select on one of the horizontal lines on the circle and with the mouse button pressed move the mouse sideways. To rotate the parts about the horizontal axis, select on one of the vertical lines on the circle and with the mouse button pressed move the mouse upward or downward.

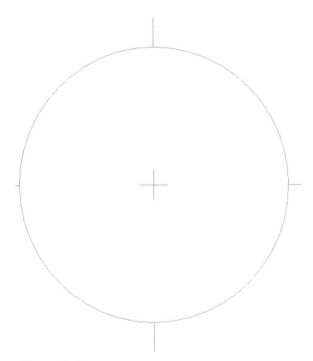

Figure 2.4

Look At

To change your viewpoint so you are looking parallel to a plane or to rotate the screen's viewpoint to be horizontal to an edge, select the Look At tool from the Standard toolbar then select a plane or edge. The Look At tool can also be issued by selecting a plane or edge then right clicking while the mouse is in the graphics area and select Look At from the context menu.

Common View (Glass Box)

To change your viewpoint to a specific viewpoint, you can issue the Common View tool by first issuing the Dynamic Rotate tool then either press the SPACE bar or right click and select Common view from the context menu. A cube will appear with arrows pointing at each corner and face similar to what is shown in Figure 2.5. Then select on one of the arrows of the cube and the viewpoint will rotate itself to that perspective. If you are looking at the cube from a plan view, you can select on an edge of the cube to rotate the cube 90°. To accept the view orientation, either press the ESC key or right click and select Done from the context menu. To change the default viewpoint that is used when Isometric View is selected from the context menu, select an arrow on the corner that you want to be the default isometric view and then right mouse click and select Redefine Isometric from the context menu.

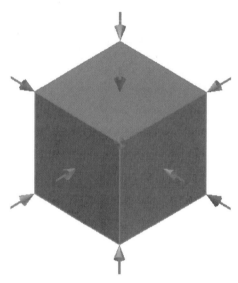

Figure 2.5

F4 Shortcut to Dynamic Rotation

While working, you can press and hold down the F4 key and the same circular image with lines at the quadrants and center will appear as shown in Figure 2.4. With the F4 key depressed, rotate the model. When you are done rotating the model, release the F4 key. If you were performing an operation, that operation will be resumed after releasing the F4 key.

Previous View

To return to the view that was previously on the screen, press the F5 key or right click in the graphics area and select previous view from the context menu. Continue this process until the view that you want to return to is on the screen.

DISPLAY OPTIONS

When working with solids there are three options for displaying the parts: Shaded Display, Hidden Edge Display, and Wireframe Display as shown in Figure 2.6 from the Standard toolbar. You can choose which mode works best for you and you can switch between the modes as you see fit. Each display mode will be described next.

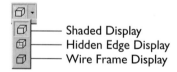

Figure 2.6

Shaded Display

The part(s) will be shaded in the color or material that was assigned to that part. Parts or faces that are behind another part or face will not be displayed.

Hidden Edge Display

The part(s) will be shaded and the edges that are behind other parts or faces will be displayed. In complex parts and assemblies this display can be confusing.

Wireframe Display

In wireframe display the part(s) will not be shaded, only the outline of the part(s) will be displayed.

Tip: You can work with and edit parts in shaded, hidden edge, or wireframe mode. Use your Intelli-Mouse wheel to zoom and pan the geometry on the screen.

TUTORIAL 2.1—VIEWING A MODEL

1. Open the file \Inventor Book\Clutch Gear.ipt
2. With the mouse in the graphics area, right mouse click and select Isometric View from the context menu.
3. Dynamically rotate the parts by using the Dynamic Rotate tool from the Standard toolbar. Rotate the part by selecting inside the circle and moving the mouse. Rotate the part by selecting outside the circle and moving the mouse. Rotate the part by selecting on one of the horizontal lines on the edge of the circle and move the mouse. Rotate the part by selecting on one of the vertical lines on the edge of the circle and move the mouse.
4. While still in the Dynamic Rotate tool, press the SPACE bar and select on different arrows.

Note: If you wanted to redefine the default isometric view, do the following. With the Glass box still on the screen, right mouse click and select Redefine Isometric. Then press the ESC key to exit the tool and accept the new viewpoint.

5. Issue the Look At tool and select a planar face, then select an edge. Practice selecting different faces and edges then press the ESC key to exit the tool and accept the new viewpoint.
6. Press and hold down the F4 key and rotate the part; lift up the F4 key to exit the tool.
7. Practice zooming and panning using the F3 and F2 keys, and the zoom and pan tools from the Standard toolbar. If you have a wheel mouse, practice zooming and panning by rolling the wheel in and out and then by pressing the wheel down and moving the mouse.
8. Alternate between shaded, hidden edge display, and wireframe modes and rotate the model dynamically in each.
9. Press the ESC KEY to exit the current tool.
10. When you feel comfortable rotating the part, close the file without saving it.

UNDERSTANDING WHAT A FEATURE IS

In the first chapter, you created a sketch, constrained, and dimensioned it. The next step is to turn the sketch into a 3-D part. The first sketch of a part that is turned into a solid is referred to as the *base* feature. There are also *sketched* features, where you draw a sketch on a planar face or work plane and either add or subtract material from an existing part. Extruding, revolving, sweeping, or lofting can create these sketched features on a part. You can also create *placed* features such as fillets, chamfers, and holes by placing them on the base feature. Placed features will be covered in chapter 3. Features are the building blocks that help create a part. For example, a plate with a hole in it would have a base feature representing the plate and a hole feature representing the hole. As the features are added to the part, they will appear in the Browser, showing the history of the part or assembly (the order in which the features were created or the parts were assembled). Features can be edited, deleted, or reordered from the part as required. The next few chapters will go through many methods of adding and deleting features.

USING THE BROWSER FOR CREATING AND EDITING

The Autodesk Inventor Browser by default is docked along the left side of the screen and displays the history of the file. In the browser you can help create, edit, rename, copy, delete, and reorder features and parts. You can expand or collapse the history of the part(s) (the order in which the features and parts were created) by selecting the + and – on the left side of the part name in the browser. Also by right clicking while in the browser you can expand or collapse all children by selecting that option from the context menu. Figure 2.7 shows the Browser with all the features expanded. As parts grow in complexity, so will the information found in the Browser. Dependent features will be indented to show that they are related to the top level. This is referred to as a parent-child relationship. The child cannot exist without the parent and is dependent on the parent. For example, if a hole is created in an extruded rectangle and the extrusion is then deleted, the hole will also be deleted. To help filter out some of the object types that appear in the browser, you can select the icon that looks like a funnel from the top of the browser. After selecting the funnel, you can select objects to hide from the drop list. Each feature in the browser is given a default name. For example, the first extrusion would be named Extrusion1 and the number will sequence as you add like features. The browser can also help you locate parts and features in the graphics area. To highlight a feature or part in the graphics area, simply move the mouse over the feature or part name in the browser and it will be highlighted. To zoom in on a selected feature, right click on the fea-

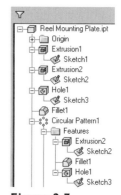

Figure 2.7

ture's name in the browser and select Find in Window from the context menu. The browser itself acts like a toolbar, except that it can be resized while docked. To close the browser, select the "X" in the upper right corner of the browser. If the browser is not visible on the screen, you can display it by selecting Browser Bar from the View pull-down menu or right click while over a toolbar and select Browser Bar from the drop list.

Specific functionality of the browser will be covered throughout the book in the sections where it pertains. However, a basic rule is to either right click or double click on the feature's name to edit or perform a function on the feature.

SWITCHING ENVIRONMENTS

To this point you have been working in the sketching environment that allowed us to work in 2-D. The next step is to turn the sketch into a feature. To turn a part into a feature, you need to exit the sketch environment and enter the feature environment. There are multiple methods to accomplish this.

- Select the Return icon on the left side of the Standard toolbar as shown in Figure 2.8.
- Select the arrow on the Panel Bar near Sketch and select Features from the drop list as shown in Figure 2.9.
- Right click in the graphics area and select Create feature from the context menu as shown in Figure 2.10 and then select the tool you need to create the feature. Only the tools that are applicable to the current situation will be available.
- Type in a shortcut key to start the feature tool.

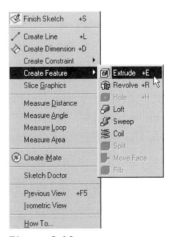

Figure 2.8   **Figure 2.9**   **Figure 2.10**

FEATURES TOOLS

Once you are in the features environment, the Panel Bar icons will change to the feature tools. Figure 2.11 shows the Features Panel Bar with the expert mode turned off. Following is a description of the feature tools.

- **Extrude:** Extrudes a sketch in the positive or negative Z-axis. This extrusion can form a base feature or add or remove material from a part. Extruding will be covered later in this chapter.

- **Revolve:** Rotates a sketch around a straight edge or axis at a specified angle. This revolution can form a base feature or add or remove material from a part. Revolving will be covered later in this chapter.

- **Hole:** Creates a drilled, tapped, countersunk, or counterbore hole feature in a part. Creation of holes will be covered in chapter 3.

- **Shell:** Removes material from the part leaving a specified wall thickness. Shelling will be covered in chapter 3.

- **Rib:** Creates a thin walled extrusion from a 2-D sketch. Creation of ribs will be covered in chapter 7.

- **Loft:** Creates a lofted base feature or loft feature by blending between sketches that lie on different planes. Lofting will be covered in chapter 7.

Figure 2.11

- **Sweep:** Creates a swept feature by sweeping a sketch about a defined path. Sweeping will be covered in chapter 7.

- **Coil:** Creates a 3-D helical part or feature by revolving a sketch around a centerline. The specifics of the 3-D helical are determined by data entered into a dialog box. Creating a coil will be covered in chapter 7.

- **Thread:** Creates a thread feature in a hole or on a cylinder. The thread will be represented in the graphic area as well as in drawing views. Creating a thread feature will be covered in chapter 3.

- **Fillet:** Creates a fillet feature on an edge or edges; the feature can be a fillet or a round. Creating fillet features will be covered in chapter 3.

- **Chamfer:** Creates a chamfer feature on a selected edge or edges. Creating chamfer features will be covered in chapter 3.

- **Face Draft:** Creates a face draft feature that angles a selected planar face in or out. Creating face draft features will be covered in chapter 3.

- **Split:** Allows a face to be split into two faces or splits a part into two. Splitting faces and parts will be covered in chapter 7.
- **View Catalog**

    View Catalog: Opens Windows Explorer to let you browse any design elements that have been created. Viewing design elements will be covered in chapter 7.

    Create iFeature: Creates a library of a feature or a part. Limits can be placed on the features. These limits can be used when the iFeature is inserted. Creating iFeatures will be covered in chapter 7.

    Insert iFeature: Place an iFeature on the current part or assembly. Inserting iFeatures will be covered in chapter 7.

- **Derived Component:** Creates a part based on another part. Any changes to the original part will also appear in the derived part. Creating derived parts will be covered in chapter 6.
- **Rectangular Pattern:** A selected feature(s) will be arrayed in a rectangular pattern. Creating rectangular patterns will be covered in chapter 3.
- **Circular Pattern:** A selected feature(s) will be arrayed in a circular pattern around a centerline. Creating circular patterns will be covered in chapter 3.
- **Mirror Feature:** A selected feature(s) will be mirrored about a plane. Mirroring features will be covered in chapter 7.
- **Work Plane:** Creates a work plane feature that is based on user input. A work plane can be used as a sketch plane or as a plane for the mirror feature tool. Creating work planes will be covered in chapter 3.
- **Work Axis:** Creates a work axis feature that can be used as an axis for a circular pattern or used to create work planes. Creating a work axis will be covered in chapter 3.
- **Work Points:** Creates a work point feature that can be used to help create work planes and other tools that require points to be selected. Creating work points will be covered in chapter 3.
- **Promote:** Will stitch an imported surface model into a single surface and then promote it so the surface can be used for creating features.

EXTRUDING THE SKETCH

The most common method for creating a part is to extrude the sketch giving the sketch depth along the Z-axis. Before extruding, it is helpful to be in an isometric view. Inventor will preview the extrusion depth and direction.

To extrude a sketch, select the Autodesk Inventor Extrude tool from the Features Panel Bar, press the hot key E, or right click in the graphics area and from the pop-up menu, select Create Feature > Extrude. After you issue the tool, the Extrude dialog box will appear as shown in Figure 2.12.

Figure 2.12

The Extrude dialog box has three sections: Shape, Operation (middle column), and Extents. When changes are made in the dialog box, the shape of the sketch will change in the graphics area to represent these values and options. When you have entered the values and the options you need, select the OK button to complete the dialog.

SHAPE

This section has two options: Profile and Taper.

Profile: Select this button to choose the sketch to extrude. If the Profile button is shown depressed as in Figure 2.12, this is telling you that a sketch or sketch area needs to be selected. If there are multiple closed profiles, you will need to select which sketch area you want to extrude. If there is only one possible profile, Inventor will select it for you and you can skip this step. If the wrong profile or sketch area is selected, reselect the Profile button and choose the new profile or sketch area.

Taper: Type the angle that you want the sketch to be drafted. A 0 draft angle extrudes the sketch straight outward, a negative number tapers the sketch inward, and a positive number expands the sketch outward (in the manufacturing industry this is known as a reverse or negative draft). An arrow will appear that represents the direction that the taper will go. A draft angle can be applied to all Extent types.

OPERATION

This is the middle column of buttons that is not labeled. If this is the first sketch that you are working with, it is referred to as a base feature and the two middle buttons will be grayed out (cut and intersect). The operation will default to Join (top button). Once the base feature has been established, you can then extrude a sketch adding or removing material from the part using the join or cut option or keep what is common between the existing part and the completed extrude operation using the intersect option.

Join: Adds material to the part.

Cut: Removes material from the part.

Intersect: Keeps what is common to the part and this second feature.

Surfaces: Extrudes the sketch and the result is a surface.

EXTENTS

The Extents area determines the type and distance the extrusion will go. This section has three areas: termination, distance, and direction.

Termination

The termination determines how the sketch will be extruded. There are five options to choose from. As in the Operation section, this section has options that are not available until a base part exists.

Distance: The sketch will be extruded a specified distance.

To Next: The sketch will be extruded until it reaches a plane or face. The sketch must be fully enclosed in the area that it is projecting to, otherwise use the "To" termination with the Extend to Surface option. Select the direction button to determine the extrusion direction.

To: The sketch will be extruded until it reaches a selected face or plane. To select a plane or face to end the extrusion, select the "Select Surface" button as shown in Figure 2.13 and then select a face or plane for the extrusion to terminate at.

From To: The extrusion will start at a selected plane or face and stop at another plane or face. To select a plane or face to start and end the extrusion, select the "Select Start Surface" button and then select a face or plane for the extrusion to start at. Then select the "Select End Surface" button and select a face or plane for the extrusion to terminate at.

All: The sketch will be extruded all the way through the part in one direction.

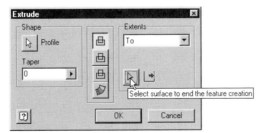

Figure 2.13

Distance

If Distance was selected as the termination, type in a value that the sketch should be extruded, or select the arrow to the right and measure two points to determine a value, or select from the list of the most recent values that were used. After a value is typed in, a preview image will appear in the graphics area representing how the extrusion will look. Another method is to select the edge of the extrusion shown in the graphics area and drag it. A preview image will appear in the graphics area, and the corresponding value will appear in the distance area. If values and units appear red as they're being entered, that means that something is incorrect and needs to be corrected. For example, if two decimal places were typed in (2.12.5) or mmm was entered for the unit, it would appear in red and would need to be corrected before the extrusion could be completed.

Direction

There are three buttons to choose from for determining the direction. Choose the first two to flip the extrusion direction or select the last button to have the extrusion go equal distances in the negative and positive directions. For example, if the extrusion distance is 2", the extrusion will go 1" in both the negative and positive Z directions.

 Note: When you convert a sketch into a part or feature, the dimensions on that sketch will disappear. When the feature is edited, the dimensions will reappear. The dimensions will also appear when drawing views are made. For more information on editing parts or features, see "Editing 3-D Parts" later in this chapter.

To select a sketch or to edit the selected sketch, select the Profile button and select the sketch.

To extend the draft angle out from the part, give the draft angle a positive number also known as reverse draft.

When extruding a sketch a given distance, drag the sketch to get a preview of how it will look at different distances.

TUTORIAL 2.2—EXTRUDING A SKETCH

In this tutorial you will practice creating a base feature with the join operation. Tutorials covering the cut and intersect operation will be covered later in this chapter.

1. Open the following file: \Inventor Book\TU1-3.ipt
2. Change to an isometric view by selecting the right mouse button while in the graphics area and then select Isometric View from the context menu.
3. Start the Extrude tool from the Features Panel Bar and the Extrude dialog box will appear.

4. Note, since there is only one possible sketch, the profile has automatically been selected.

5. In the Extrude dialog box, change the Distance to 10mm. The preview of the extruded sketch should change to reflect this value.

6. Select the edge of the extrusion that is shown in the graphics area and drag it to a distance of 1.25. Your screen should look similar to Figure 2.14. When the value is 1.25, remove your finger from the mouse button.

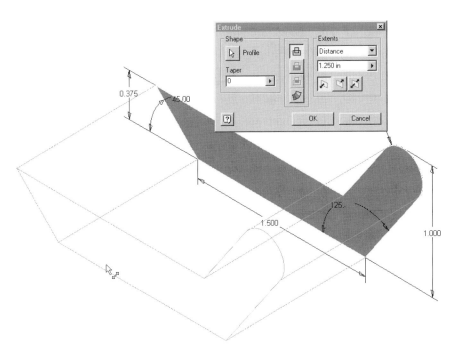

Figure 2.14

7. Adjust the value of the Taper to 10 and then select the F4 key and rotate the part until you can see the arrow showing that the taper will go toward the outside. Then release the F4 key.

8. Change the direction of the extrusion by selecting the middle button on the direction area. The extrusion direction should flip to show the direction away from the profile.

9. Change the direction to Midplane by selecting the right button on the direction area.

10. Change back to an isometric view by selecting the right mouse button while in the graphics area and then select Isometric View from the context menu. Your screen should resemble Figure 2.15.

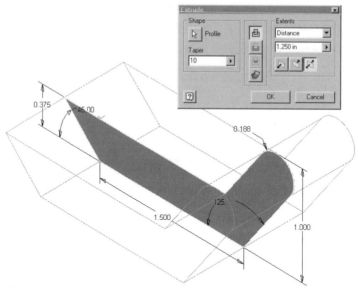

Figure 2.15

11. Select the OK button to complete the dialog and create the extrusion. When you are done, your screen should resemble Figure 2.16.

12. Select Save Copy As from the File pull down menu and save it with the following name: \Inventor Book\TU2-2.ipt

13. Close all files.

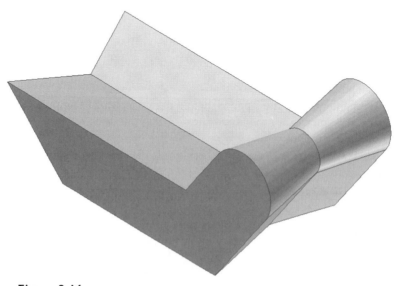

Figure 2.16

REVOLVING THE SKETCH

Another method for creating a part is to revolve a sketch around a straight edge or axis (centerline). Revolve can be used to create cylindrical parts or features. To revolve a sketch, you will follow the same steps you did to extrude the sketch (create the sketch, add constraints and dimensions), then select the Revolve tool from the Features Panel Bar, press R and ENTER, or right click in the graphics area and from the pop-up menu, select Create Feature > Revolve. The Revolve dialog box will appear as shown in Figure 2.17. The Revolve dialog box has three sections: Shape, Operation (middle column) and Extents. When changes are made in the dialog box, the shape of the sketch will change in the graphics area to represent these values and options. When you have entered the values and the options you need, select the OK button to complete the operation.

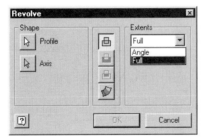

Figure 2.17

SHAPE

This section has two options: Profile and Axis.

Profile: Select this button to choose the profile to revolve. If the Profile button is shown depressed, this is telling you that a profile or sketch area needs to be selected. If there are multiple closed profiles, you will need to select the profile you want to revolve. If there is only one possible profile, Inventor will select it for you and you can skip this step. If the wrong profile or sketch area is selected, reselect the Profile button and choose the new profile or sketch area.

Axis: Select a straight edge or centerline that the sketch should be revolved about. The edge does not need to be part of the sketch. If a centerline is used, it needs to be part of the sketch. See the section below on how to create a centerline and create diametric dimensions.

OPERATION

This is the middle column of buttons that is not labeled. If this is the first sketch that you are working with, it is referred to as a base feature, and the two middle buttons will be grayed out (cut and intersect). The operation will default to Join (top button). Once the base feature has been established, you can then revolve a sketch, adding or removing material from the part using the join or cut options, or you may keep what is common between the existing part and the completed revolve feature using the intersect option.

Join: Adds material to the part.

Cut: Removes material from the part.

Intersect: Keeps what is common to the part and this second feature.

Surfaces: Revolves the profile and the result is a surface.

EXTENTS

The Extents area determines if the sketch will be revolved 360° or a specified angle.

Full: Full is the default option and will revolve the sketch 360° about a specified edge or axis.

Angle: Select this option from the drop list, and the Revolve dialog box will give you more options as shown in Figure 2.18. Enter an angle that the sketch should be revolved. There are three buttons below the degree area that will determine the direction of the revolve. Choose the first two to flip the revolve direction or select the last button to have the revolve go equal distances in the negative and positive directions. For example, if the angle was set to 90°, the revolve will go 45° in both the negative and positive directions.

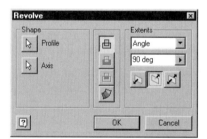

Figure 2.18

CENTERLINES AND DIAMETRIC DIMENSIONS

When revolving a sketch, the majority of the time you will want to specify diameter dimensions instead of radial dimensions. Sketches that are revolved usually are a quarter section of the completed part. Figure 2.19 shows a sketch that represents a quarter section of the completed part with a centerline and the diameter dimensions. The dimensions that are placed on the sketch will be used for the drawing views. If you want to place a diameter dimension on the sketch, a centerline must be selected.

To create a centerline, first draw a line that will become the centerline. Then exit the line tool and select the line, then from the Style drop list on the Command Bar select Centerline as shown in Figure 2.20. The line will then become a centerline. To create a diameter dimension, use the General Dimension tool and pick either the centerline and the other point or line to be dimensioned or select a point or edge and then the centerline to place the diameter dimension. When selecting the centerline make certain to select the entire centerline not just an endpoint of the centerline.

Tip: When creating diametric dimensions, select the entire centerline. If a point is selected from the centerline, a radial dimension will be created.

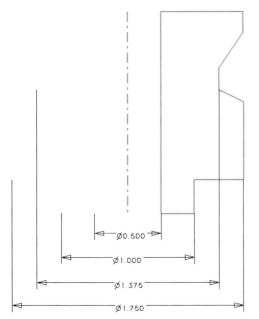

Figure 2.19

Figure 2.20

TUTORIAL 2.3—REVOLVING A SKETCH

1. Start a new part file based on the default Standard.ipt file.
2. Sketch the geometry on the right side of the centerline as shown in Figure 2.21.
3. Draw the left vertical line that will become the centerline.
4. Exit the line tool and select the line that you just drew and pick the Centerline from the Style drop list as shown in Figure 2.18.

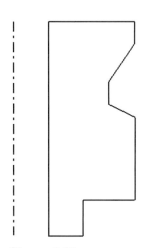

Figure 2.21 **Figure 2.22**

Ø0.500
Ø1.000
Ø1.375
Ø1.750

5. Add the diametric dimensions as shown in Figure 2.22.

6. Change to an isometric view by selecting the right mouse button while in the graphics area and then select Isometric View from the context menu.

7. Select the Revolve tool from the Features Panel Bar and then the Revolve dialog box will appear.

8. In the dialog box, change the extents to Angle and enter 45 degrees. When done, the graphics area will change to reflect this value.

9. Change the direction of the revolve by picking the middle flip button. The preview image will reverse the direction.

10. Change the revolve direction to Midplane and the degrees to 90, and then your screen should resemble Figure 2.23.

11. Change the Extent type to Full and then select the OK button to complete the operation. When finished, your screen should resemble Figure 2.24.

12. Save the file with the following name: \Inventor Book\TU2-3.ipt

13. Close the file.

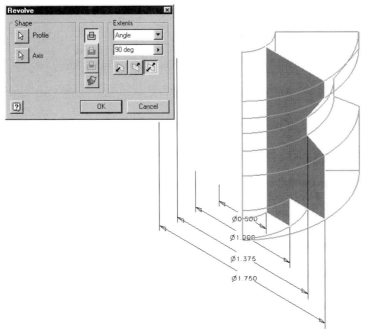

Figure 2.23

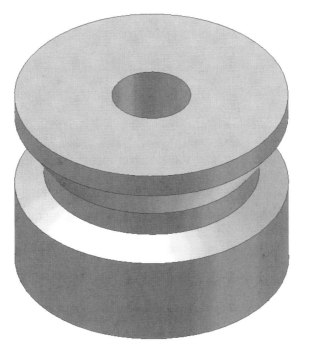

Figure 2.24

EDITING A FEATURE (FEATURE EDITING)

After the feature was created, any dimension that was visible in the sketch was consumed into the feature. If you need to change the dimensions, values, taper, operation, or termination type that was entered through the dialog box, you will need to edit the sketch or feature. To edit the information that was entered through the dialog box while the feature was created, follow these steps.

- Right click on the feature's name in the browser and choose Edit Feature from the context menu. Figure 2.25 shows the context menu that appears after right clicking on Extrusion1.
- In the dialog box, type in new values or change the settings.
- Pick the OK button to complete the edit.

Everything can be changed in the dialog box except the Join operation on a base feature.

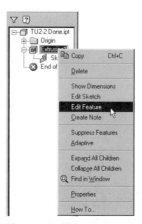

Figure 2.25

To change the dimensional values of a feature, you need to edit the feature so the dimensions are visible. There are multiple methods that can be used to edit the dimensions. There is no preferred method; you may choose any method you want to edit a features dimensions.

- In the Browser, double click on the feature's name.
- In the Browser, right click on the feature's name and choose Edit Sketch from the pop-up menu that appears.
- In the Browser, expand the children of the feature and then right click on the name of the sketch and choose Edit Sketch from the context menu as shown in Figure 2.26.

- From the Command Bar, select the down arrow on the Select button and choose Feature Priority from the drop list as shown in Figure 2.27. Alternatively, you can hold down the SHIFT key and with the mouse in the graphics area, right mouse click and select Feature Priority from the context menu. Then double click on the feature that you want to edit and the dimensions will appear on the part.

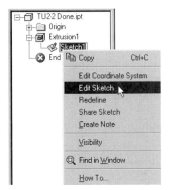

Figure 2.26

Figure 2.27

After the dimensions are visible on the screen, double click on the dimension text that you want to edit. The Edit Dimension dialog box that was used to create the original dimension will appear. Type in a new value and then select the check mark in the dialog box or press the ENTER key. Continue to edit the dimensions and when complete, select the Return, Sketch, or the Update button from the Command Bar as shown in Figure 2.28. The dimensions will disappear and the new values will be used to regenerate the part.

Figure 2.28

FAILED FEATURES

If the browser turns red after updating the part, the new values or settings were not successfully regenerated. You MUST pick the Undo button from the Standard toolbar until the part reappears correctly. Then edit the feature and type in new values or select different settings.

RENAMING FEATURES AND SKETCHES

By default each feature is given a name. These feature names may not help you when trying to edit a specific feature of a complex part. The feature names will not be descriptive to your design intent. For example, the first extrusion by default will be given the name "Extrusion1," whereas the design intent may be that the extrusion is the thickness of a plate. All feature names can be edited. To rename a feature, select the feature name with a slow double click. Then type in a new name;

spaces are allowed. For our example, the first extrusion named "Extrusion1" will be renamed to "Plate Thickness."

FEATURE COLOR

When parts become complex, you may want to change colors of specific features. To change a feature color, right click on the feature name in the Browser and select Properties from the context menu. In the Feature Properties dialog, select a new feature color from the drop list as shown in Figure 2.29, and then select OK to complete the operation. The feature's name can also be changed in the top area of the Feature Properties dialog box.

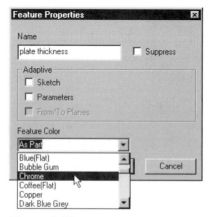

Figure 2.29

DELETING A FEATURE

You may choose to delete a feature after it has been placed. To delete a feature, right click on the feature name in the browser and select Delete from the context menu as shown in Figure 2.30. Then the Delete Features dialog box will appear; from the list, choose what should be deleted. Multiple features can be deleted by holding down the CTRL key and selecting their names in the browser and then select Delete from the context menu.

Figure 2.30

TUTORIAL 2.4—EDITING FEATURES

1. Open the following file: \Inventor Book\TU2-2.ipt
2. Change to an isometric view by selecting the right mouse button while in the graphics area and then select Isometric View from the context menu.
3. Rename the feature "Extrusion1" to "Base Extrusion."
4. Edit the feature by right clicking on "Base Extrusion" in the Browser and select Edit Feature from the context menu.
5. Change the Taper to "0 degrees," the distance to "1," and the direction to point out toward the screen. Figure 2.31 shows the filled-in dialog box and the preview image. When done, select OK to complete the modifications.

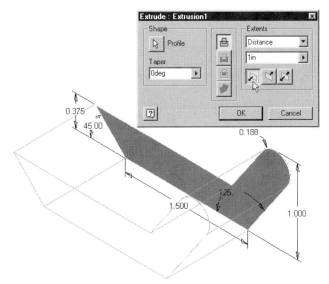

Figure 2.31

6. In the Browser double click on the feature "Base Extrusion." Change some dimension values by double clicking on the text and entering new values. When done, select the Update button from the command bar.
7. In the Browser right click on the feature "Base Extrusion" and select Edit Sketch from the context menu. Change some dimension values by double clicking on the text and entering new values. When done, select the Update button from the command bar.
8. From the Command Bar change the Select option to Feature Priority and then double click on the base feature in the graphics area. Change some dimension values by double clicking on the text and entering new values. When done, select the Update button from the command bar.

9. Then in the browser right click on the feature "Base Extrusion" and select Properties from the context menu. From the Feature Color drop list, select a color of your choice. Select OK to complete the modification.

10. Close the file without saving the changes.

EDITING A FEATURE'S SKETCH

In the last section you learned how to edit the dimensions and the settings in which the feature was created. In this section you will learn how to add and delete constraints, dimensions, or geometry to the original 2-D sketch. To edit the 2-D sketch of a feature, do the following.

- In the Browser right click on the features name or the name of the sketch that you want to edit and choose Edit Sketch from the context menu like what is shown in Figure 2.32.

- Then add or remove geometry from the sketch and add and delete constraints and dimensions as you learned in chapter 1.

While editing the sketch you can both add and remove objects to the sketch. When adding geometry to the sketch, lines, arcs, circles, and splines can be added. Objects can also be deleted. To delete an object(s) select it and either right click and select Delete from the context menu or once the objects are selected press the DELETE key. If you erase an object from the sketch that has dimensions associated to it, the dimensions are no longer valid for the sketch and will be erased. The entire sketch can also be erased and replaced with an entirely new sketch. When replacing entire sketches, if there are other features that would be consumed by the new objects, they should be deleted first and recreated. Once the sketch has been modified, update the part by selecting the Return, Sketch, or the Update button from the Command Bar.

 Note: If you get an error after updating the part, make sure that the sketch forms a closed profile. If the appended sketch forms multiple closed profiles, you will need to reselect the profile area.

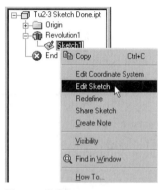

Figure 2.32

TUTORIAL 2.5—EDITING A FEATURE'S SKETCH

1. Start a new part file based on the default Standard.ipt file.
2. Sketch the geometry and add the constraints and dimensions as shown in Figure 2.33. The figure is shown with all the constraints visible. Note: Fix the lower left corner.
3. Change to an isometric view by selecting the right mouse button while in the graphics area and then select Isometric View from the context menu.
4. Select the Extrude tool and then extrude the profile a distance of 1 inch.
5. Edit the feature by right clicking on "Extrusion1" in the browser and select Edit Sketch from the context menu.
6. In the sketch delete the arc.
7. Delete the parallel constraint on the right vertical line.
8. Draw a horizontal line to close the gap.
9. If needed add a perpendicular constraint between the top horizontal and right vertical line.
10. Add a 60 degree angular dimension between the bottom horizontal and right vertical line.
11. Then add a 1 inch horizontal dimension to the bottom horizontal line. When you are finished your screen should look similar to Figure 2.34, shown with all the constraints visible and in an isometric view.
12. Update the part and the sketch will be returned to a 3-D part.
13. Practice editing the sketch by adding and removing constraints, dimensions, and objects to the sketch. When you feel comfortable editing sketches, close the file without saving it.

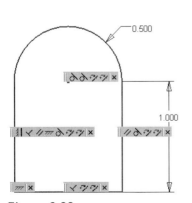

Figure 2.33

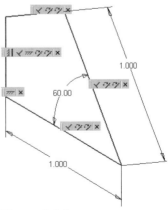

Figure 2.34

SKETCHED FEATURES

A sketched feature is a feature that you draw on a plane and add or remove material from the part. Follow these basic steps to create a sketched feature.

- Create or make an existing sketch active.
- Draw the geometry that defines the sketch.
- Add constraints and dimensions.
- Perform a Boolean operation that will either add or remove material from the part or keep whatever is common between the part and the completed feature.

There are no limits to the number of sketched features that can be added to a part. However, each sketched feature needs to be on its own plane. Even if multiple sketched features will exist on the same plane of the part, a new sketch will need to be created for each sketched feature. In this section you will learn the final step—how to assign a plane to the active sketch.

ASSIGNING A PLANE TO THE ACTIVE SKETCH

As was stated in the last section, each sketch must exist on its own plane. The active sketch has a plane on which the sketch is drawn. There are three requirements to assign a plane to the active sketch.

- The part in which the plane will be placed must be an Inventor part.
- The part must be active.
- A plane that the sketch will be created on must be a planar face or a work plane. The planar face does not need to have a straight edge. For example, a cylinder has two faces, one on the top and the other on the bottom of the part; neither has a straight edge, but a sketch can be placed on either face.

To make a sketch active, follow one of the following methods.

- Start the New Sketch tool from the Command Bar as shown in Figure 2.35 and then select the plane where you want to place the sketch.

Figure 2.35

- Press the hot key s and then select the plane where you want to place the sketch.
- Select a plane that will contain the active sketch and issue the New Sketch tool from the Command Bar.
- Select a plane that will contain the active sketch and press the hot key s.
- Select a plane that will contain the active sketch and right click in the graphics area and from the context menu select New Sketch.

- Start the New Sketch tool from the Command Bar, expand the Origin folder in the Browser, and select one of the default work planes.
- Expand the Origin folder in the Browser, right click on one of the default work planes, and from the context menu select New Sketch.
- To make a sketch that has already been created the active sketch, start the New Sketch tool and then select the sketch name from the browser.

Once a sketch is created it will appear in the browser with the name Sketch#, and sketch tools will appear in the Panel Bar. The number will sequence for each new sketch that is created. In the browser these sketches can be renamed by double clicking slowly on the name and typing a new name. After a new sketch has been created on a plane, it is sometimes easier to work in a plan view (looking straight at the current plane). This can be done by using the Look At tool and then selecting the plane. Now that the sketch has been created, you can place sketch entities and apply constraints and dimensions exactly as you did with the first sketch. In addition to constraining and dimensioning the new sketch, you can also constrain the new sketch to the existing part. You can place dimensions to geometry that does not lie on the current plane; however, the dimensions will be placed on the current plane. When you look at a part from different viewpoints, you will see arcs and circular edges appearing as lines. Remember that because they are still circular edges, any constraints and dimensions you add will go to their center points. After constraining and dimensioning the sketch, the sketch can be extruded, revolved, or swept. The sweep and loft commands will be explained in chapter 7. To delete a sketch, first exit the sketch environment by selecting the sketch name in the browser and then either right clicking with the mouse in the graphics area and selecting Finish Sketch, or selecting the Sketch button in the Command Bar. There can only be one active sketch at a time.

FACE CYCLING

Inventor has dynamic face highlighting to help you select the correct face to make the active sketch. As you move the mouse over a given face it will highlight. The active plane will be shown with a black edge; the highlighted face will be shown in red. Keep moving the mouse, and different faces will highlight as the mouse passes over them. To cycle to a face that is behind another one, move the mouse over the face that is in front of the one that you want to select. Hold the mouse still and an image will appear like what is shown in Figure 2.36. Pick the left or right arrow to cycle through the faces until the correct face is highlighted and then press the left mouse button or pick the green rectangle in the middle of the image. If you have a wheel mouse, you can cycle through the faces by spinning the wheel.

Figure 2.36

Tip: To reactivate a sketch, select its name in the browser and make it active using the New Sketch tool.

To see the sketch in a plan view, use the Look At tool and select the plane or select the plane and then select the Look At tool.

Rename the sketches to better explain what they are being used for.

There can only be one active sketch at a time.

TUTORIAL 2.6—SKETCH PLANES

1. Start a new part file based on the default Standard.ipt file.
2. Draw a square and place two 3-inch dimensions on it.
3. Change to an isometric view by selecting the right mouse button while in the graphics area and then select Isometric View from the context menu.
4. Extrude the sketch 2 inches in the default direction with –15° of taper.
5. Make a new sketch on each of the planar faces and draw a circle on each plane. Try to select planes that are below another by cycling through the part. Also try to rotate the part while in the New Sketch tool to see the new plane from a better point of view. **Tip:** Press and hold down the F4 key to dynamically rotate the part. When complete, your screen should resemble Figure 2.37 shown in a wireframe display with the content of the browser.
6. Rotate the part to verify that there are circles on each of the planes.
7. Save the file with the following name: \Inventor Book\TU2-6.ipt.
8. Close the file.

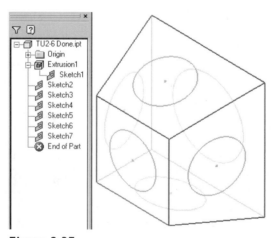

Figure 2.37

TUTORIAL 2.7—SKETCH PLANES

1. Start a new part file based on the default Standard.ipt file.
2. Draw and dimension a 2-inch diameter circle.
3. Change to an isometric view by selecting the right mouse button while in the graphics area and then select Isometric View from the context menu.
4. Extrude the sketch 3 inches in the default direction with no taper.
5. Create a new active sketch on the front planar face of the cylinder.
6. Draw a square in the middle of this cylindrical face.
7. Create a new active sketch on the back planar face of the cylinder.
8. Draw a triangle in the middle of this cylindrical face.
9. Rotate the part to verify that the square and triangle are on the correct planes.
10. When complete, your screen should resemble Figure 2.38.
11. Save the file with the following name: \Inventor Book\TU2-7.ipt.
12. Close the file.

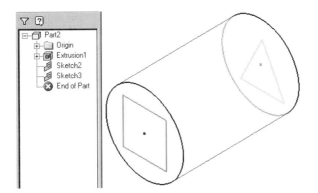

Figure 2.38

CUT, JOIN, AND INTERSECT OPERATION

The next step after the active sketch has been defined and geometry has been drawn is to add constraints and dimensions. Then use that sketch to add or remove material to the part or keep the common material between the part and the finished feature. When working with the sketch there are three operations that can be used: join, cut, or intersect. The tools that you use to add these sketched features are extrude, revolve, sweep, or loft. There are four basic steps that you will follow when creating sketched features.

1. Create a New Sketch on a planar face or work plane.
2. Sketch the geometry.
3. Constrain and dimension the sketch as needed.
4. Then extrude, revolve, sweep, or loft the sketch adding or removing material from the part or keeping what is common to the part and the finished feature.

After selecting the Extrude, Revolve, Sweep, or Loft tool, a dialog box will appear and you can then select the Operation and Termination type. Descriptions follow for both the Operation and Termination selections that will be needed for extruding, revolving, sweeping, and lofting.

OPERATION

There are four options to choose from.

Join: Adds material to the part.

Cut: Removes material from the part.

Intersect: Keeps what is common to the part and this feature.

Surfaces: Extrudes the sketch and the result is a surface. Surfaces can be used as a cutting edge or as a face to extend to. The use of surfaces will be covered in chapter 7.

Figure 2.39 shows a pictorial representation of how the join, cut, and intersect operations work. The top row represents a part on the left side and a completed feature on the right side. The bottom row shows the part after the new feature is created.

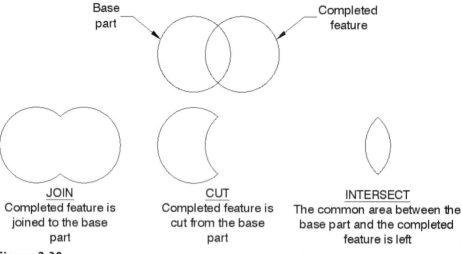

Figure 2.39

TERMINATION

The operation determines how the sketch will be extruded. There are five options to choose from. As in the Operation section, this section has options that are not available until a base part exists.

Distance: The sketch will be extruded a specified distance.

To Next: The sketch will be extruded until it reaches a plane or face. The sketch must be fully enclosed in the area that it is being projected to, otherwise use the "To" termination with the Extend to Surface option. Select the direction button to determine the extrusion direction.

To: The sketch will be extruded until it reaches a selected face or plane. To select a plane or face to end the extrusion, use the "Select Surface" button as shown in Figure 2.13 and then select a face or plane for the extrusion to terminate at.

From To: The extrusion will start at a selected plane or face and stop at another plane or face. To select a plane or face to start and end the extrusion, pick the "Select Start Surface" button and then select a face or plane for the extrusion to start at. Then pick the "Select End Surface" button and select a face or plane for the extrusion to terminate at.

All: The sketch will be extruded all the way through the part in one direction.

Tip: There are no limits to the number of features that a part can have.

All features are displayed in the browser in the order in which they are created unless they are reordered.

When creating a part, think about what the finished part will look like and create the part one feature at a time.

Press and hold down the F4 key to quickly rotate a part(s) to another viewpoint.

TUTORIAL 2.8—OPERATIONS AND TERMINATIONS

1. Start a new part file based on the default Standard.ipt file.
2. Draw and dimension a 2-inch by 1-inch rectangle.
3. Change to an isometric view by selecting the right mouse button while in the graphics area and then select Isometric View from the context menu.
4. Extrude the rectangle .1 inch.
5. Create a new sketch on the bottom of the extrusion as shown in Figure 2.40.

Figure 2.40

6. Use the Look At tool and select the sketch that you just created.
7. Draw and dimension the sketch shown in Figure 2.41. The arc is *not* tangent to the top or bottom line.

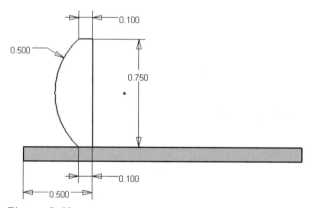

Figure 2.41

8. Change to an isometric view by selecting the right mouse button while in the graphics area and then select Isometric View from the context menu.
9. Extrude the sketch up 1 inch and join material; before extruding the sketch your screen should resemble Figure 2.42.

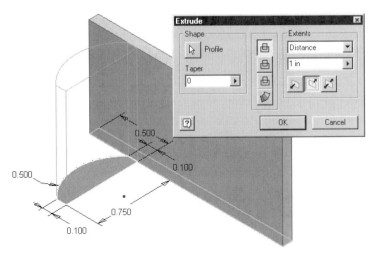

Figure 2.42

10. Create a new sketch on the bottom of the extrusion as shown in Figure 2.40.
11. Draw and dimension the sketch shown in Figure 2.43. The arc is tangent to the bottom line.

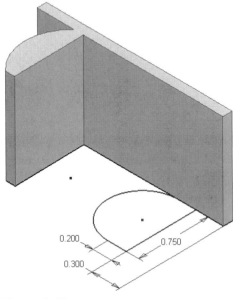

Figure 2.43

12. Extrude the sketch using the "To" termination and use the join operation. You will need to select the "Select Surface to End" button and select the top plane of the part. Since the sketch is not fully enclosed in the plane, place a check mark in the "Extend to Surface" button. Before extruding the sketch your screen should resemble Figure 2.44.

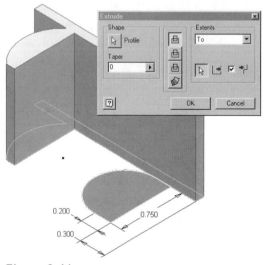

Figure 2.44

13. Edit the first extrusion and change the dimensions from 2 to 3 inches and the 1 to 1.5 inches; then Update your part. When done your screen should look similar to Figure 2.45. Extrusion2 will stay at 1 inch and Extrusion3 will

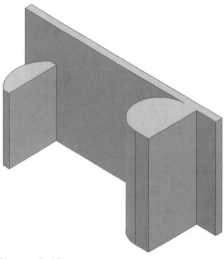

Figure 2.45

be extended to 1.5 inches because its extrusion termination was linked to the plane that changed its value.

14. Create a new sketch plane on the back small plane of the part as shown in Figure 2.46.

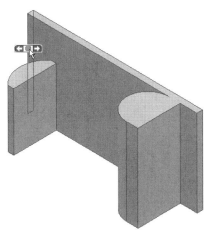

Figure 2.46

15. Use the Look At tool, then select the sketch that you just created. Orientate the view so that the rectangle depth of .1 is vertical. If the view is rotated issue the Look At command and select the line that you want to be horizontal. Your screen should look similar to Figure 2.47 shown in wireframe display.

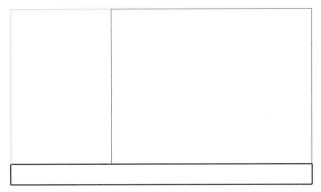

Figure 2.47

16. Draw and dimension the sketch shown in Figure 2.48.

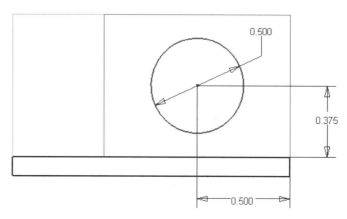

Figure 2.48

17. Change to an isometric. Use the Dynamic Rotate tool and press the space bar to access Common View (Glass Box). Select the arrows until your viewpoint matches Figure 2.49.

18. Extrude the sketch using the To Next termination and use the Join operation. Before extruding the sketch, your screen should resemble Figure 2.49. When the extrusion is complete your screen should resemble Figure 2.50.

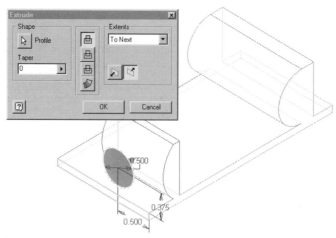

Figure 2.49

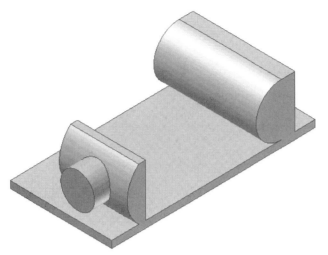

Figure 2.50

19. Edit the last extrusion and change the termination to "From To." Select the first circular face as the surface to start at and select the furthest circular face as the surface to end at. Before extruding the sketch, your screen should resemble Figure 2.51. The extend buttons do not need to be selected since the sketch will be fully enclosed by the selected faces. When the extrusion is complete, your screen should resemble Figure 2.52.

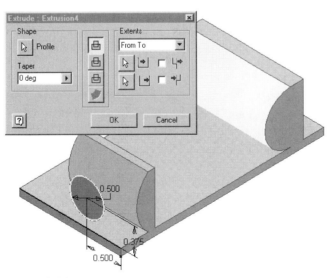

Figure 2.51

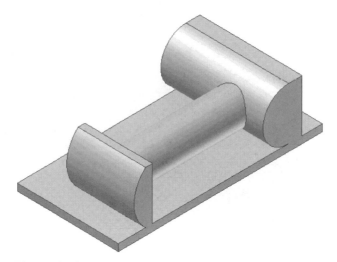

Figure 2.52

20. Create a new sketch on the small plane of the part as shown in Figure 2.53; for clarity it is displayed in wireframe.

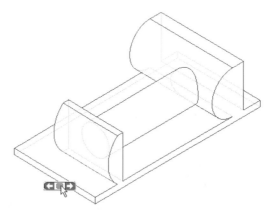

Figure 2.53

21. Draw and dimension the rectangle as shown in Figure 2.54. The bottom of the rectangle should be coincident to the bottom edge of the part.

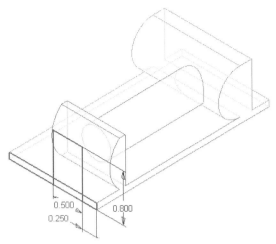

Figure 2.54

22. Extrude the sketch using the Cut operation and "All" as the termination. Before extruding the sketch, your screen should resemble Figure 2.55; for clarity the image is displayed in wireframe. When the operation is complete your screen should resemble Figure 2.56.

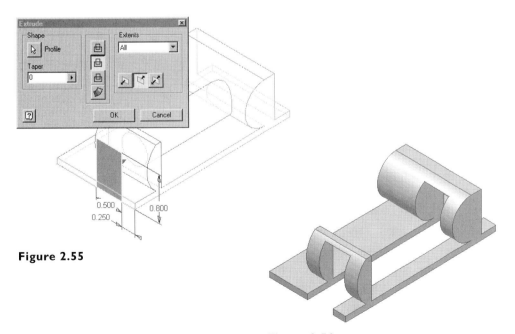

Figure 2.55

Figure 2.56

23. Edit the last extrusion and change the Operation to "Intersect." Before extruding the sketch, your screen should resemble Figure 2.57. When the command is complete, your screen should resemble Figure 2.58.

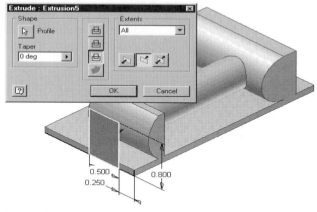

Figure 2.57

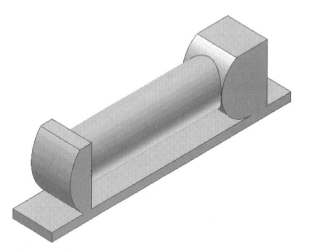

Figure 2.58

24. Practice adding different types of sketched features. When you feel comfortable adding sketched features, save the file with the following name: \Inventor Book\TU2-8.ipt
25. Close the file.

Viewing, Extruding, Revolving, Editing Features, Sketched Features, and Boolean Operations 85

PRACTICE EXERCISES

The following exercises are intended to challenge you by providing problems that are open-ended. As with the kinds of problems you'll encounter in work situations, there are multiple ways to arrive at the intended solution. In the end, your solution should match the sketch that is shown. Follow the directions for each of the exercises. When you have completed the part save the file.

Exercise 2.1—Bracket

Open the file \Inventor Book\Bracket.ipt. Then create the part based on the information in Figure 2.59.

Tip: Extrude the sketch .25 inches, select the profile in-between the slot and the exterior sketch boundary; this will form the center island. Then add the other two-sketched features. When done save the file.

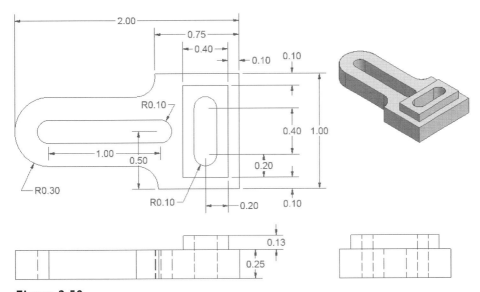

Figure 2.59

Exercise 2.2—Connecting Rod

Open the file \Inventor Book\Connecting Rod.ipt. Then create the part based on the information in Figure 2.60.

Tip: Extrude the sketch .25 inches with the midplane option add −3° of draft. When done save the file.

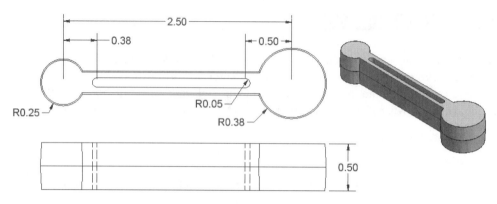

NOTE: NEGATIVE 3 DEGREES OF DRAFT ON OUTSIDE EXTRUSION

Figure 2.60

Exercise 2.3—Cylinder

Start a new file based on the Standard.ipt template. Create the part based on the information in Figure 2.61.

 Tip: Sketch a quarter profile, create a centerline and then place diametric dimensions and revolve the sketch.

When done modeling save the finished part as \Inventor Book\Cylinder.ipt

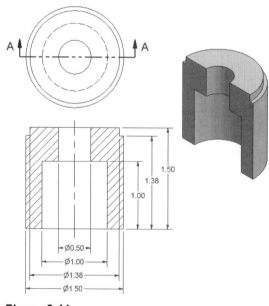

Figure 2.61

REVIEW QUESTIONS

1. Press and hold down the F5 key to dynamically rotate a part(s). True or False?
2. Explain how to redefine the default isometric view.
3. The Look At tool will change the viewpoint to an isometric view. True or False?
4. A part can only be edited while it is in shaded display. True or False?
5. What is a base feature?
6. When creating a feature with the Extrude or Revolve tool, you can drag the sketch to define the distance or angle. True or False?
7. What objects can be used as an axis of revolution?
8. Explain how to create a diametric dimension on a sketch.
9. Name two ways to edit an existing feature.
10. Once a sketch becomes a base feature, you cannot delete or add constraints, dimensions, or objects to the sketch. True or False?
11. What is the active sketch?
12. Each sketch must exist on its own plane. True or False?
13. Name three operation types used to create sketched features?
14. A Cut operation cannot be performed before a base feature (part) is first created. True or False?
15. Once a sketched feature exists its termination cannot be changed. True or False?

CHAPTER 3

Placed Features

In chapter 2 you learned how to create base and sketched features. In this chapter you will learn how to create placed features. Placed features are features that are predefined except for specific values and only need to be located. Placed features can be edited from the Browser like sketched features. When a placed feature is edited, either a dialog box that was used to create it will context or feature values will appear on the part. When creating your part, it is usually better to use placed features instead of sketched features wherever possible. For example, to make a through hole you could draw a circle profile and dimension it, then extrude it using the All extension option. Alternatively, you could select the type of hole, size, and then place it through a dialog box. Then when drawing views are generated, the type and size of the hole can be easily annotated and will be automatically updated if the hole type or values change.

CHAPTER OBJECTIVES

After completing this chapter, you will be able to:

- create fillets.
- create chamfers.
- create holes.
- create internal and external threads.
- shell a part.
- add face draft to a part.
- create work axes.
- create work planes.
- create work points.
- pattern features.

FILLETS

When creating fillets in 3-D, you will select the edge that needs to be filleted and the fillet will be created between the two faces sharing this edge. This differs from filleting in 2-D because in 2-D you select two objects and a fillet is created between them. When creating a part, it is usually good practice to create fillets and chamfers as one of the last features in the part. Fillets add complexity to the part, which in turn adds to the file size and removes edges that may be needed to place other features.

To create a fillet feature, select the Fillet tool from the Feature Panel Bar as shown in Figure 3.1. After you select the tool, the Fillet dialog box appears as shown in Figure 3.2. The Fillet dialog box has three tabs: Constant, Variable, and Setbacks; each tab will create a different type or style of fillet. The options for each of the tabs will be described in the following section. Before we look at the tabs let's look at the methodology that will be used to create fillets. You can either select an edge or edges to fillet or select the type of fillet to create. If an edge was selected before the fillet feature tool was issued, it will be put into the first selection set. Inventor places its fillets into selection sets. Selection sets contain the selected edges. Each selection set can contain its own fillet value. There is no limit to the number of selection sets that can exist on a part. However, an edge can only exist in one selection set. All the selection sets will be shown as a single fillet feature in the browser. To add to the first selection set, pick the edges that you want to have the same radius. To create another selection set, pick in the "Click here to add" area and then select the edges that will be part of the next selection set. To remove an edge that has been selected, pick the selection set that the edge is part of and the edges will highlight. Then hold down the CTRL key and select the edge to be removed from the selection set. After selecting the edge(s) to fillet, in the Fillet dialog box select the type

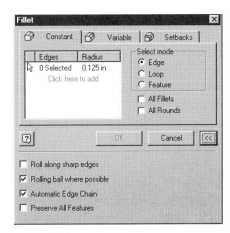

Figure 3.1   Figure 3.2

of fillet that you want to create and then enter the values for the fillet. As changes are made in the dialog box, a representation of the fillet will be previewed in the graphics area. When the fillet type and value are correct, select the OK button to create the fillet.

To edit a fillet's type and radius follow these guidelines.

- Start the Edit Feature tool by right mouse clicking on the fillets name in the browser and select Edit Feature from the context menu.
- The Fillet dialog box will appear with all the settings that were used to create it. Then change the fillets settings as needed.
- To edit only the dimensional value of a fillet you can double click on the fillet's name in the browser or change the select priority to Feature Priority and then double click on the fillet that is on the part.
- The dimension(s) will appear on the part. Double click on the dimension to change it and type in new values through the Edit Dimension dialog box. Then select the check mark in the dialog box or enter.
- Pick the Update tool to update the part.

CONSTANT TAB

With the options under the Constant tab, as shown in Figure 3.2, you will create fillets that have the same radius from the beginning to the end of the fillet. There is no limit to the number of edges that can be filleted with a constant fillet. The edges can be selected as a single set or selected in multiple sets; each set can have its own radius value. The order in which the edges are selected is not important. The edges that are to be filleted need to be picked individually, the use of window or crossing selection method is not allowed. If you change the value of a group, all the fillets in that group will change. To remove an edge from a group, pick the group in the select area and then hold down the CTRL key and select the edge. Below is a description of the options that are available from the Constant tab.

Select Edge and Radius: By default, after starting the fillet tool, you can select edges and they will appear in the first selection set. To create another selection set, click in the "Click here to add" area and then select the edges that will be part of the next selection set. After picking an edge, a preview image of the fillet will appear on the edge that reflects the current values.

Select Mode
Edge: Select the Edge mode to pick individual edges to fillet. By default any edge that is tangent to the selected edges will be selected. If you do not want to have tangent edges automatically be selected, uncheck the Automatic Edge Chain option in the more section.

Loop: Select the Loop option to have all the edges that form a closed loop to the selected edge be filleted.

Feature: The Feature option will select all the edges of a selected feature.

All Fillets: The All Fillets option will select all concave edges of a part that have not already been filleted. The All Fillets option adds material to the part and requires a separate edge selection set.

All Rounds: The All Rounds option will select all convex edges of a part that have not already been filleted. The All Rounds option removes material from the part and requires a separate edge selection set.

More Options

Roll Along Sharp Edges: When selected, a constant radius will be maintained and adjacent faces will be extended as needed.

Rolling Ball Where Possible: When selected, a fillet around a corner will be created that looks like a ball has been rolled along the edges that define the corners.

Automatic Edge Chain: When selected, tangent edges will be automatically selected after picking an edge.

Preserve All Features: When selected, all features that intersect with the fillet are checked and their intersections are calculated during the fillet operation. If the check box is cleared, only the edges that are part of the fillet operation are calculated during the operation.

VARIABLE TAB

With the options under the Variable tab as shown in Figure 3.3, you will create a fillet that has a different starting and ending radius.

Edges: By default, after selecting the fillet tool, you can select an edge and it will appear in the first selection set. For a variable fillet, only one edge can exist per selection set. After picking an edge, a preview image of the fillet will appear on the edge that reflects the current values. To add another point, select on the edge and its name will appear in the Point column.

Point: To determine where the start or end point is on the selected edge, pick the Start or End text and a filled cyan circle will appear at the point on the edge to represent it.

Radius: After selecting a points name, type in a radius.

Position: If another point was added, type in value between .000001 and 1. This number represents a percentage between the start and end point relative from the start point.

Smooth Radius Transition: The fillet will blend from the starting to the ending radius as a smooth transition, like a cubic spline. If unchecked the fillet will blend from the starting to the ending radius in a straight line.

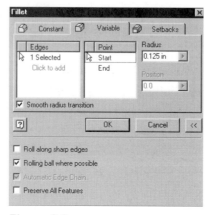

Figure 3.3

SETBACKS

With the options under the Setback tab, as shown in Figure 3.4, you will specify the distance that a fillet will start its transition from its vertex. Setbacks can only be used where three filleted edges form a vertex.

Vertex: Select a vertex on the part that has intersecting edges.

Edge: Select an edge that intersects the selected vertex. After selecting the edge, you can type in a setback value.

Setback: After selecting an edge, type in a distance that the radius will start its transition. Each edge can have its own setback value.

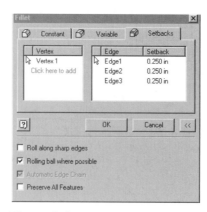

Figure 3.4

Placed Features 93

 Tip: If you get an error when creating or editing a fillet, try to create it with a smaller radius. If you still get an error after trying to use a smaller fillet, you can try to create the fillets in a different sequence or to create multiple fillets in the same operation.

TUTORIAL 3.1—CREATING FILLETS

1. Open the file \Inventor Book\Fillet Features.ipt
2. Select the Fillet tool from the Features Panel Bar and select the top edges of the small extrusion. Type in a radius of .0625 as shown in Figure 3.5.

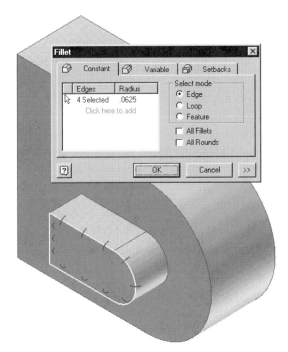

Figure 3.5

3. Pick in the Click here to add area and then select the bottom edges of the small extrusion and type in a radius of .125 as shown in Figure 3.6. Then select OK to create the fillets.

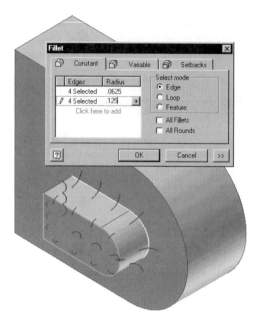

Figure 3.6

4. Start the Fillet Feature tool and select the All Fillets option and type in a radius of .25 and then select the All Rounds option and type in a radius of .0625 as shown in Figure 3.7. Then select OK to create the fillets.

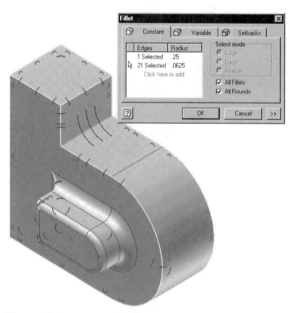

Figure 3.7

5. Edit the feature "Fillet2" and change the radius of the first selection set (All Fillets) to .125 and then select OK to update the feature. When done your screen should resemble Figure 3.8.

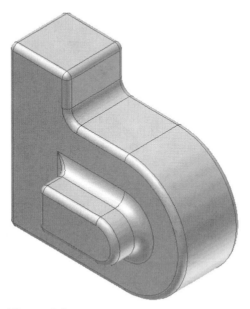

Figure 3.8

6. Delete the feature "Fillet2" by right clicking on its name in the browser and select Delete from the context menu.
7. Start the Fillet Feature tool and pick the Variable tab. Then select the edge as shown in Figure 3.9. Type in the value of .0625 for the Start and End points.
8. Next you will add a point to define the radius for the middle of the fillet. While still in the fillet operation, select near the middle of the same edge. Type in a radius of .125 and a position of .5.
9. While still in the fillet operation, click in the Click here to add area of the Fillet dialog box and then select an adjacent edge that lies on the top plane and use the same settings and procedures that you did for the first edge with a start and end radius of .0625 and a point in the middle of the edge and a radius value of .125. Do this for the remaining top two edges. When done, select the OK button and your screen should resemble Figure 3.10.

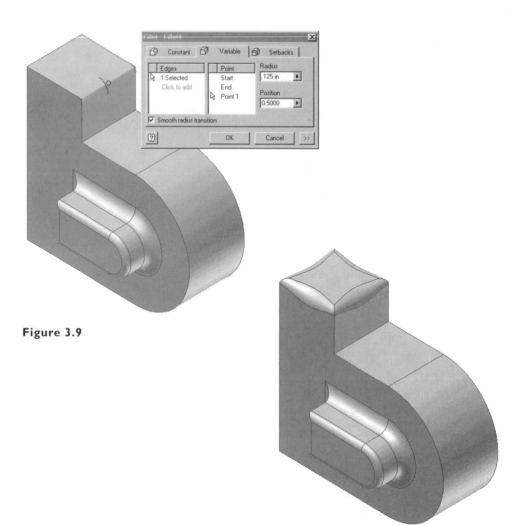

Figure 3.9

Figure 3.10

10. Practice creating and editing fillets with different settings that are available from the More area. When you feel comfortable with fillets, close the file without saving the changes.

CHAMFERS

To create a chamfer feature, you will follow the same steps as you did to create the fillet features. Select the common edge and the chamfer will be created between the two faces sharing this edge. To create a chamfer feature, select the Chamfer tool from the Feature Panel Bar as shown in Figure 3.11. After you select the tool, the

Chamfer dialog box appears as shown in Figure 3.12. The Chamfer dialog box has four areas: Methods, Edges, Distance, and More. Each will be described. As with fillet features there are selection sets and all the selection sets will be represented in the browser as a single feature. From the dialog box, select a method, type in a Distance or angle, select the edge or edges to chamfer, and then select OK to create the chamfer.

To edit the type of chamfer feature or distances, perform one of the following methods.

- Double click on the features name in the browser.
- Right mouse clicking on the chamfer's name in the browser and select Edit Feature from the context menu.

Figure 3.11

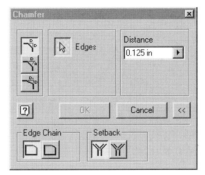

Figure 3.12

- Change the select priority to Feature Priority and then double click on the chamfer on the part. The Chamfer dialog box will appear with all the settings that were used to create it. Change the settings as needed.

METHOD

Distance: The distance option will create a 45° chamfer on the selected edge. The size of the chamfer is determined by typing in a distance in the dialog box. The value is then offset from the two common faces. A single edge, multiple edges, or a chain of edges can be selected in a single selection set. A preview image of the chamfer will appear on the part. If the wrong edge was picked, select on the Edge button and choose a new edge.

Distance and Angle: The Distance and Angle option will create a chamfer offset from a selected edge on a specified face, at an angle from the number of degrees specified. In the dialog box type an angle and distance that the chamfer will be. Then select the face that the angle will be based on and specify an edge to be chamfered.

One or multiple edges can be selected; the edges must lie on the selected face. A preview image of the chamfer will appear on the part. If the wrong face or edge was picked, select on the Edge or Face button and then choose a new face or edge.

Two Distances: The Two Distances option will create a chamfer offset from two faces, each the amount that you specify. First, select an edge and type a value for Distance1 and Distance2. A preview image of the chamfer will appear. To reverse the direction of the distances pick the Flip button. When the correct information about the chamfer is in the dialog box select the OK button to create the chamfer. Only a single edge or chained edges can be used with the two-distance option.

EDGE AND FACE

Edge: Select an edge or edges to be chamfered.

Face: Select a face that the chamfer will be based from.

Flip: Select the button to reverse the direction of the distances for the Two Distance chamfer.

DISTANCE AND ANGLE

Distance: Type in a distance that will be used for the offset.

Angle: Type in a value that will be used for the angle if creating the Distance Angle chamfer type.

EDGE CHAIN AND SETBACK

Edge Chain: Select if tangent edges will automatically be selected after picking an edge.

Setback: When the distance method is used and three chamfers meet at a vertex you can choose to have the intersection of the three chamfers form a flat edge (left button) or the intersection can meet at a point as though the edges were milled (right button).

TUTORIAL 3.2—CREATING CHAMFERS

1. Start a new part file based on the default Standard.ipt file.
2. Sketch and dimension a 2-inch square.
3. Change to an isometric view by selecting the right mouse button while in the graphics area and then select Isometric View from the context menu.
4. Extrude the sketch 1 inch.
5. Select the Chamfer feature tool from the Features Panel Bar. Use the distance option and select the four edges as shown in Figure 3.13, and change the distance to .25 inches and select OK to create the chamfer.

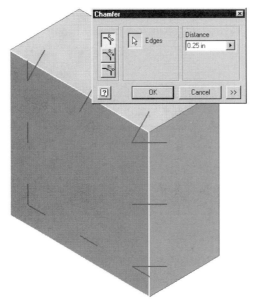

Figure 3.13

6. Edit the chamfer that you just created and remove the left and bottom edges by holding down the CTRL key and selecting them. Then add the top right edge as shown in Figure 3.14 and select OK to create the chamfer.

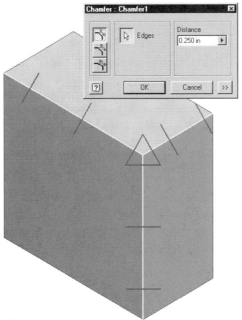

Figure 3.14

7. Edit the chamfer and change the setback to No Setback as shown in Figure 3.15. Then select OK to create the chamfer. When done your screen will resemble Figure 3.16.

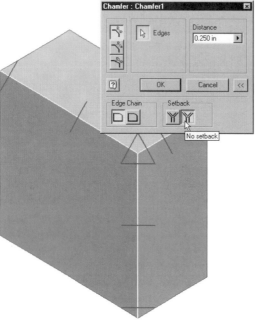

Figure 3.15

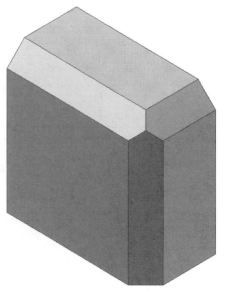

Figure 3.16

8. Delete the feature Chamfer1.
9. Pick the Chamfer tool. Use the Distance and Angle option and select the left most vertical face, and the front face should become highlighted. Then select the four surrounding edges and in the Chamfer dialog box change the distance to .3 inches and angle to 60 degrees and then your screen should resemble Figure 3.17. Pick the OK button to create the chamfer.

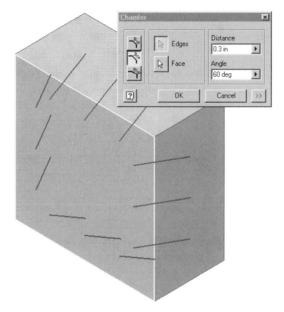

Figure 3.17

10. Delete the chamfer.
11. Pick the Chamfer tool. Use the Two Distances option and select an edge of your choice. Type in different values for Distance1 and Distance2 and then select the Flip button to reverse the order of the distances on the edge. When done, select OK to create the chamfer.
12. Practice creating and editing chamfers. When you feel comfortable with chamfers, close the file without saving the changes.

HOLES

There are four basic types of holes that Inventor can create: drilled, tapped, counterbore, and countersink holes. To create a hole feature, follow these guidelines.

- Before creating a hole, create a new sketch that the hole will exist on.
- After activating the sketch, Hole Centers will be created to represent the center of the hole(s). If the hole is going to be located from two edges, you

- will need to place a Hole Center(s) on the active sketch. If the hole(s) that you are going to create will be concentric to a circular edge that lies on the active sketch, hole centers will automatically be created at the center points or the circular edges. Hole centers will be covered in the next section.
- After the Hole Center(s) have been placed, select the Hole tool from the Feature Panel Bar as shown in Figure 3.18 or press the hot key H.

Figure 3.18

- The Holes dialog box will appear that has four tabs containing options for creating holes. Each of the four tabs will be explained below. The Centers button is depressed by default; select any Hole Centers where you want to create a hole.

Note: Any endpoint of a line or sketch curves that are on the active sketch can also be used as Hole Centers. If multiple Hole Centers are selected, they will be grouped together as a single feature in the browser. To deselect a Hole Center, hold down the CTRL key and select it. After the Hole Centers have been selected, pick the options that you need from the Holes dialog box. As the options are changed, a preview image of the hole(s) will appear on the part. When you are done making changes, select the OK button to create the hole(s).

To edit the type of Hole feature or distances perform one of the following methods.

- Double click on the features name in the browser.
- Right mouse click on the Hole's name in the browser and select Edit Feature from the context menu.
- Change the select priority to Feature Priority and then double click on the hole feature on the part in the graphics window.

After starting the hole feature, the Holes dialog box will appear with all the settings that were used to create it. Change the settings as needed.

Type Tab

Under the Type tab, shown in Figure 3.19, you will establish the type of hole and its termination.

Centers: Select the hole center point(s) where you want to create a hole.

Hole Type: Select the type of hole that you want to create: drilled, counterbore, or countersink.

Termination: Select how the hole(s) will terminate.

> **Distance:** Specify a distance for the depth of the hole.
>
> **Through All:** The hole(s) will extend through the entire part in one direction.
>
> **To:** Select a plane that the hole will stop at.
>
> **Flip:** Reverse the direction that the hole will travel.

Dimensions: To change the diameter or depth of the hole(s), select the dimension in the dialog box and type in a new value.

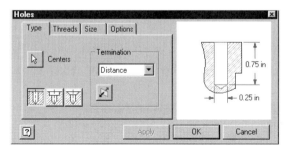

Figure 3.19

Threads Tab

Under the Threads tab, shown in Figure 3.20, you will determine if a hole will be tapped and if so set its values.

Tapped: Select this box if the hole(s) will be tapped.

Full Depth: Select this box if the threads will extend the full depth of the hole.

Thread Type: From the drop list select the standard that the hole(s) will be based on: ANSI Unified Screw Thread or ANSI Metric M Profile.

Left Hand: Select this option to create threads that wind counter clockwise and recede.

Right Hand: Select this option to create threads that wind clockwise and recede.

Dimensions: To change the depth of the thread(s), select the dimension in the dialog box and type in a new value.

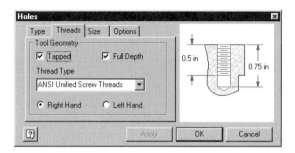

Figure 3.20

Size Tab

Under the Size tab, shown in Figure 3.21, you will set the nominal size for the thread.

Nominal Size: Select the nominal size for the thread from the drop list.

Pitch: Select the pitch size for the thread from the drop list.

Class: Select the class of thread from the drop list.

Diameter: Select how you want to define the diameter of the thread from the drop list.

Dimensions: To change the depth of the thread(s), select the dimension in the dialog box and type in a new value.

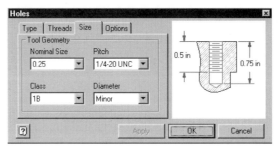

Figure 3.21

Options Tab

Under the Options tab as shown in Figure 3.22, you will define the angle for countersink holes and angle for the tip of the drill point.

Drill Point: Select if the drill should be flat or angled and adjust the angle as needed.

Countersink Angle: If the hole type has been set to countersink, you can set its angle here.

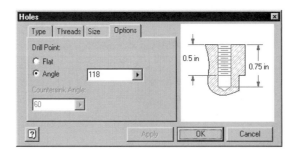

Figure 3.22

HOLE CENTERS

Hole Centers are placed features that are used to create holes that will lie on a plane or are used as points that can be sketched to. To create a Hole Center follow these guidelines.

- Make an active sketch.
- Pick the Point, Hole Center tool from the Sketch Panel Bar as shown in Figure 3.23.
- Pick a point where the hole will be placed; or move the cursor over objects until a coincident constraint glyph appears; or move the cursor over a line and it will snap to the endpoint or midpoint; or right mouse click and select Midpoint, Center, or Intersection from the context menu to snap to the desired point.

When you place a "Point, Hole Center," the default style will be a Hole Center. To change the style to a Sketch Point, select Sketch Point style from the style drop list as shown in Figure 3.24. The style can be changed before or after the point is placed by selecting the Point or Hole Center and then changing its style from the drop list. Once the Hole Center(s) is placed, you can add constraints and dimensions as needed.

Figure 3.23

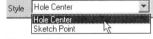

Figure 3.24

TUTORIAL 3.3—CREATING HOLES

1. Open the file \Inventor Book\Hole Features.ipt
2. Make a New Sketch on the front face of the extruded cylinder.
3. Pick the Hole tool from the Features Panel Bar and then select the center point of the cylindrical edge. In the dialog box, change the depth to .5 inches and the diameter to .3125 inches as shown in Figure 3.25. Select OK to create the hole.

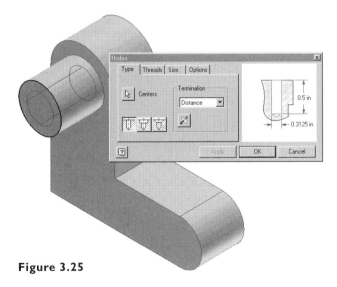

Figure 3.25

4. Make a New Sketch on the front face of the first extrusion.
5. Pick the Hole tool from the Features Panel Bar and then select the center point of the right cylindrical edge. In the dialog box, change the hole type to Counterbore; change the termination to Through All; change the drill diameter to .125; and the counterbore diameter depth and diameter to .25 in as shown in Figure 3.26. Select OK to create the hole.

Figure 3.26

6. Make a New Sketch on the front face of the first extrusion.
7. Pick the Point, Hole Center tool from the Sketch Panel Bar and place a hole center near the lower left corner of the face and then constrain the point by placing two .375 dimensions from the corner to the Hole Center as shown in Figure 3.27 (for clarity the figure is shown in wireframe display).

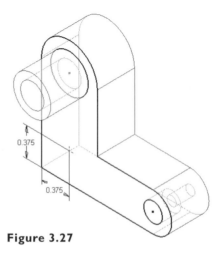

Figure 3.27

Placed Features 107

8. Pick the Hole tool from the Features Panel Bar. The Hole Center that you just created is automatically selected. Change the termination to Through All and the Hole type to Drilled. From the Threads tab check Tapped and Full Depth. From the Size tab change the Nominal size to .5 and the Pitch to 1/2-13 UNC as shown in Figure 3.28. Then select OK to create the tapped hole as shown in Figure 3.29. (Threads will be covered in the next section.)

9. Close the file without saving the changes.

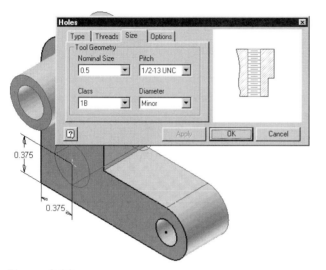

Figure 3.28

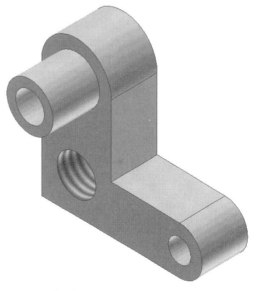

Figure 3.29

THREADS

Thread features can be used to create both internal and external threads. The threads will appear on the parts with a graphical representation. When a drawing view is created, the thread can be called out per drafting standards. Since the threads are graphical representations, they do not physically exist on the part, if a model were to be cast directly from the part, no threads would exist on the finished model. A thread feature can be added to any hole, internal or external cylinder. If you are adding threads to a hole, it can be done using the Hole tool or through the Thread tool. If it is added using the Thread tool, it will be a separate feature in the browser. For that reason, it is recommended to use the Hole tool with the tapped option when creating a tapped hole.

To create external threads, follow these guidelines.

- Create a cylinder and dimension it to the size that will represent the major diameter of the thread (this value must lie between the maximum and minimum major diameter).
- To create the Thread Feature, start the Thread tool from the Feature Panel Bar shown in Figure 3.30.
- Enter the data into the Thread Feature dialog box as needed.

Figure 3.30

There are two tabs in the Thread Feature dialog; each will be described. The thread data comes from an Excel spreadsheet named Thread.xls. By default the spreadsheet is located in the Program Files\Autodesk\Inventor <*version*>\Design Data directory. This spreadsheet can be modified to match your company's standards. This thread data is used to display the thread on the part and when the thread is annotated in a drawing view. The thread data is not associative to the threads on existing parts. When changes are made to the database, the new values will only be used when new threads are created. If a dimensional change is made to the diameter of the cylinders where a thread has been placed, a warning dialog box will appear when the part is updated. The dialog box will tell you that an inappropriate thread size was used. Accept the warning message and then edit the thread feature to a size that will fit the corresponding diameter.

LOCATION TAB (See Figure 3.31)

Face: Select this button and then select a cylindrical or conical face to place a thread on or in.

Display in Model: When checked, a visual representation of the thread will appear on the part; if unchecked, the thread will only appear when a drawing view is created.

Thread Length

Full Length: When checked, the thread will continue the entire length of the selected face.

Flip: Select this button to reverse the direction of the thread.

Length: When Full Length is not checked, enter a value for the length of the thread.

Offset: When Full Length is not checked, enter a value that the thread will be offset. The offset distance will be from the closest plane in relation to where you selected the face.

SPECIFICATION TAB (See Figure 3.32)

Thread Type: From the drop list, select the type; the types are defined in the Thread.xls file.

Nominal Size: From the drop list, select the nominal size for the thread.

Pitch: From the drop list select the pitch size for the thread.

Class: From the drop list, select the class of thread.

Right Hand or Left Hand: Select the direction for the thread, thread size, and its representation. The representation does not change by selecting the other direction.

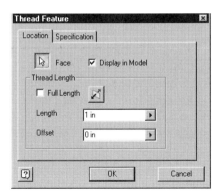

Figure 3.31

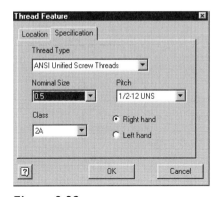

Figure 3.32

TUTORIAL 3.4—CREATING THREADS

1. Start a new part file based on the default Standard.ipt file.

2. Sketch and dimension two circles that are concentric to each other and have diameters of 1 inch and .5 inches.

3. Change to an isometric view by selecting the right mouse button while in the graphics area and then select Isometric View from the context menu.

4. Extrude the sketch 2 inches. For the profile, select the area that is in between the two circles. When complete you will have a cylinder.

5. Pick the Thread tool from the Features Panel Bar and select the inside face of the cylinder, then from the Specification tab set the Nominal Size to .5 and the Pitch to 1/2-13 UNC as shown in Figure 3.33. Then select OK to create the thread.

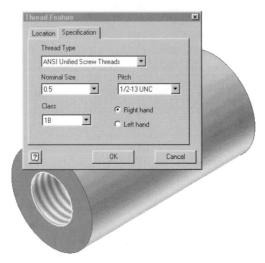

Figure 3.33

6. Pick the thread tool and select the outside face of the cylinder and uncheck Full Length and set the Length to 1 inch and the offset to .5 inches as shown in Figure 3.34. Then from the Specification tab set the Nominal Size to 1 and the Pitch to 1-8 UNC. Then select OK to create the thread.

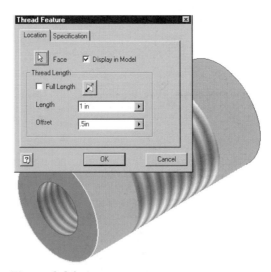

Figure 3.34

7. Edit the last Thread feature that you just created and change the thread length to Full Length.

8. Edit the first extrusion and change the diameter of the outside circle to .75 inches. Then Update the part. A warning dialog box will appear alerting you to a design problem as shown in Figure 3.35.

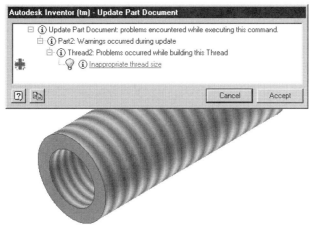

Figure 3.35

9. Accept the warning and then edit the thread feature and change its Nominal size to .875 and the pitch to 7/8-9 UNC. Select OK to complete the edit.

10. Practice creating and editing threads. When done, close the file without saving the changes.

SHELLING

While creating parts you may need to create a part that is made up of thin walls. The easiest way to create a thin walled part is to create the main shape and then use the Shell tool to remove material from the part. The term shell refers to giving the outside shape of a part a thickness (wall thickness) and removing the remaining material, like scooping out the inside of a part, leaving the walls a specified thickness. The wall thickness can be offset in, out, or evenly in both directions. If the part that is shelled contains a void like a hole, that feature will have the thickness built around it. A part may contain more than one shelled feature and individual faces of the part can have different thickness. If a wall has a different thickness than the shell thickness, it is referred to as a unique face thickness. If a face that you select for a unique face thickness has faces that are tangent to it, those faces will also have the same thickness. Faces can be removed from being shelled, and these faces will be left open. If no face is removed, the part will be hollowed out on the inside.

To create a shell feature, follow these guidelines.

- Create a part that will be shelled.
- Pick the Shell tool from the Feature Panel Bar as shown in Figure 3.36.
- The Shell dialog box will appear as shown in Figure 3.37. Enter the data as needed.
- After filling in the information in the dialog box, select OK and the part will be shelled.

The following section will explain the options that are available for the Shell tool.

To edit a shell feature, perform one of the following methods.

- Double click on the feature's name in the browser.
- Right mouse click on the name of the shell feature in the browser and select Edit Feature from the context menu.
- Change the select priority to Feature Priority and then double click the shell feature on the part.

The Shell dialog box will appear with all the settings that were used to create the feature. Change the settings as needed.

Figure 3.36

Figure 3.37

REMOVE FACES

Pick the Remove Faces button and then select the face or faces that will be left open. To deselect a face, pick the Remove Faces button and hold down the CTRL key while you select the face.

THICKNESS

Type in a value or select a previously used value from the drop list to be used for the shell thickness.

DIRECTION

Inside: Offsets the wall thickness by the given value into the part.

Outside: Offsets the wall thickness by the given value out of the part.

Midplane: Offsets the wall thickness evenly into and out of the part by the given value. A midplane offset cannot have overrides.

UNIQUE FACE THICKNESS

This is available by selecting the >> button that is located on the lower right corner of the dialog box.

To give a specific face a thickness, click in the "Click here to add" area and then select the face and type in a value. A part may contain multiple faces that have a unique thickness.

TUTORIAL 3.5—SHELLING A PART

1. Open the file \Inventor Book\Shell Features.ipt
2. Pick the Shell tool from the Features Panel Bar and pick the Remove Faces button. Then select the left front face and type in a Thickness of .03 as shown in Figure 3.38.

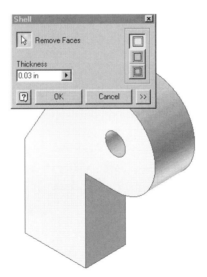

Figure 3.38

3. Select the More button >>, then click in the "Click here to add" area and then select the bottom face and then type in a value of .25 as shown in Figure 3.39. When done, select OK to create the shell.

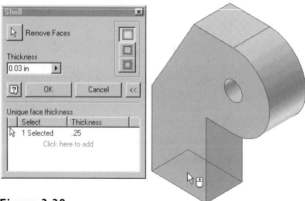

Figure 3.39

4. Change to an isometric view that gives you the viewpoint that resembles Figure 3.40.

5. Pick the Shell tool and choose the Remove Face button and then select the left front face and type in a Thickness of .03 as shown in Figure 3.40. When done, select OK to create the shell and then your screen should resemble Figure 3.41.

6. Practice deleting and creating different shell features. When done, close the file without saving changes.

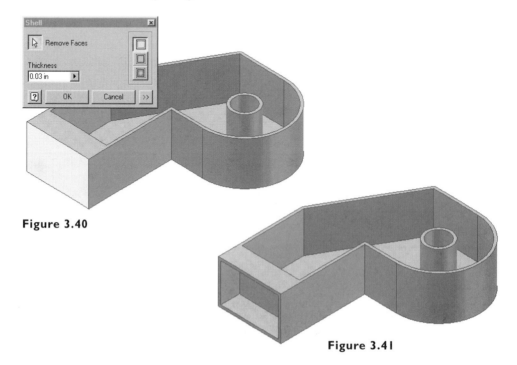

Figure 3.40

Figure 3.41

FACE DRAFT

Face draft is a feature that applies an angle to a face. Face draft can be applied to any specified internal or external face including shelled parts. When face draft is applied, any tangent face will also have the face draft applied to it. The face draft feature will be labeled "TaperEdge" in the browser.

To create a face draft feature, follow these guidelines.

- Start the Face Draft tool from the Feature Panel Bar as shown in Figure 3.42.
- A Face Draft dialog box will appear as shown in Figure 3.43. The options in the Face Draft dialog box will be explained in the next section.
- First select the pull direction that shows how the mold will be pulled out from the part.
- Then select a face or faces to have the face draft applied to. If an incorrect face is selected, it can be deselected by holding down the CTRL key and selecting the face.

Figure 3.42

Figure 3.43

PULL DIRECTION

Select the direction that the mold will be pulled out from the part. The angle will expand out in this direction. When the correct direction is displayed on the screen, press the left mouse button and an arrow will appear that shows the direction that the mold will be pulled from the part.

Direction: Select the arrow button and then move the mouse around the part. A dashed line will appear—this line is pointing 90° from the highlighted face and shows the direction that the mold will be pulled out. When the correct direction is displayed on the screen, press the left mouse button.

Flip: Select the flip button to reverse the pull direction 180°.

FACES

Select the face or faces to which the face draft will be applied. As you move the mouse over the face a symbol with an arrow will appear that shows how the draft will be applied from the nearest edge. As the mouse is moved to different edges, the arrow direction will show how the draft will be applied.

DRAFT ANGLE

Type in a value for the draft angle or select from the drop list a previously used draft angle.

TUTORIAL 3.6—APPLYING FACE DRAFT

1. Start a new part file based on the default Standard.ipt file.
2. Sketch and dimension a rectangle that measures 2 inches horizontally by 1 inch vertically.
3. Change to an isometric view by selecting the right mouse button while in the graphics area and then select Isometric View from the context menu.
4. Extrude the rectangle 3 inches.
5. Shell the part .1 inch and remove the top face.
6. Pick the Face Draft tool from the Features Panel Bar and select near the middle of the removed face, and a dashed line will appear as shown in Figure 3.44. When it appears, select the left mouse button and an arrow that points up will be shown.

Note: If the arrow is pointing in the wrong direction, picking the Flip button in the Face Draft dialog box will reverse it.

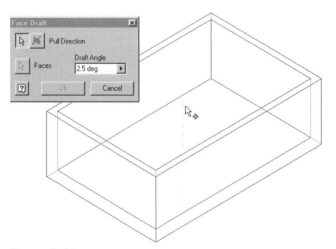

Figure 3.44

7. Change the draft angle to 3.
8. In the Face Draft dialog box, select the Faces button and pick near the inside top right edge of the box as shown in Figure 3.45. This is the edge that will be fixed.

Placed Features 117

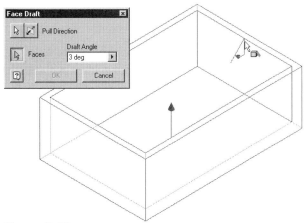

Figure 3.45

9. While still in the same operation, select near the inside back bottom edge of the box as shown in Figure 3.46. This is the edge that will be fixed. Then select the OK button to complete the Face Draft feature. Your part should resemble Figure 3.47 shown in wireframe display.

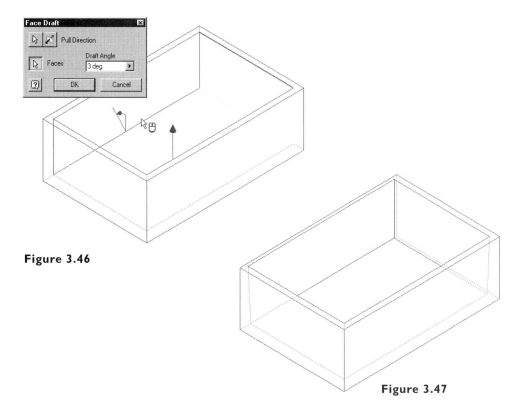

Figure 3.46

Figure 3.47

10. Delete the Taper Edge feature that you just created.
11. Place a .25 fillet in each of the four inside vertical edges of the box.
12. Pick the Face Draft tool and select near the middle of the removed face and a dashed line will appear as shown in Figure 3.44. When it appears, select the left mouse button and an arrow that points up will be shown.
13. Change the draft angle to 5.
14. In the Face Draft dialog box, select the Faces button and pick near the inside top back edge of the box as shown in Figure 3.48. Since all the faces are tangent, all the top edges that are inside the part will be highlighted. These edges will be fixed. Then select the OK button to complete the operation and your part should resemble Figure 3.49 shown in wireframe display.

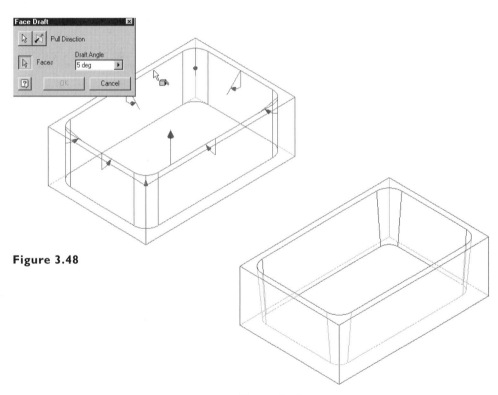

Figure 3.48

Figure 3.49

15. Practice editing the face draft by flipping the pull direction. Then create a face draft on the outside faces of the part.
16. When you are done practicing creating face drafts, close the file without saving the changes.

CREATING A WORK AXIS

A work axis is a feature that acts like a construction line that is infinite in length and can be used to: create work planes, as an axis of rotation for polar patterns and assembly constraints can be applied to them. Their length will always extend beyond the part, as the part changes size, the work axis will change so that it is always longer that the part. A work axis is parametrically tied to the part. As changes occur to the part, the work axis will maintain its relationship to the points or edge that it was created from. To create a work axis, use the Work Axis tool from the Feature Panel Bar as shown in Figure 3.50. Then use one of the following methods to create a work axis.

Figure 3.50

- Pick a cylindrical face, and a work axis will be created along its axis of revolution.
- Pick two points on a part, and a work axis will be created through both points.
- Select an edge on a part, and the work axis will be created on top of the edge.
- Pick a work point or sketch point, and a plane or face and a work axis will be created that is normal to the selected plane or face, and it will go through the point.
- Pick two nonparallel planes, and a work axis will be created at their intersection.

CREATING WORK POINTS

A work point is a feature that can be created on the active part or in 3-D space. A work point can be created any time a point is required. To create a work point, use the Work Point tool from the Feature Panel Bar as shown in Figure 3.51; or while using the Work Plane or Work Axis tool, right click and select Create Point. If the tool is issued while creating a work feature, the tool will be exited after the work point is created.

Figure 3.51

A work point can be created by doing one of the following.

- Sect an endpoint or midpoint of an edge; select an edge and axis, and a work point will be created at the intersection or theoretical intersection of the two.
- Select an edge and plane, and a work point will be created at the intersection or theoretical intersection of the two.
- Select three nonparallel faces or planes, and a work point will be created at their intersection or theoretical intersection.

TUTORIAL 3.7—CREATING A WORK AXIS AND WORK POINTS

In this exercise you will create a work axis that is based upon the location of a work point and a sketch point. You will also create a work axis based on a cylindrical face. The work axes that will be created will be used as an axis of rotation in the circular pattern exercise later in this chapter. For clarity the figures in the tutorial are displayed in wireframe.

1. Open the file \Inventor Book\Work Axis and Work Point.ipt
2. In this step you will create a work point by selecting two edges. Pick the Work Point tool from the Features Panel Bar and select the bottom left and right edges as shown in Figure 3.52.

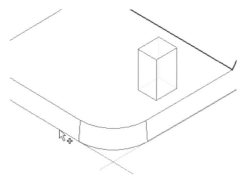

Figure 3.52

3. In this step you will create a work axis that goes through a work point and is normal to a plane. Pick the Work Axis tool from the Features Panel Bar and select the work point you just created and the inside horizontal plane as shown in Figure 3.53.

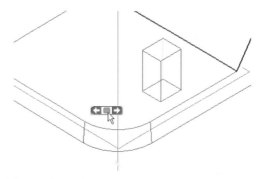

Figure 3.53

4. In this step you will create a work axis that goes through a sketch point that was already created and is normal to a plane. Pick the Work Axis tool and select the sketch point and the inside angled plane as shown in Figure 3.54.

Figure 3.54

5. In this step you will create a work axis by selecting a cylindrical face. Pick the Work Axis tool and select the outside cylindrical face shown in Figure 3.55.

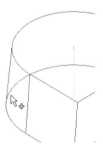

Figure 3.55

6. Save the file.

CREATING WORK PLANES

Before introducing work planes, it is important that you understand when you need to create a work plane. You can use a work plane when you need to sketch a feature and there is no planar face at the location you need to sketch. If you want a feature to terminate at a plane and there is no existing plane, you will need to create a work plane. If you want to apply an assembly constraint to a plane on a part and there is no existing plane, you will need to create a work plane. In any of these scenarios, if a plane exists you should use it and not create a work plane. A New Sketch can be created on a work plane.

A work plane is a feature that looks like a rectangular plane. It is parametrically tied to the part and will always be larger than the part. If the part moves, the work plane will also move. For example, if a work plane is tangent to the outside face of a 1″ diameter cylinder and the cylinder diameter changes to 2″, the work plane moves with the outside face of the cylinder. You can create as many work planes on a part as needed, and any work plane can be used to create a New Sketch. A work plane is a feature and must be edited and deleted like any other feature.

Before creating a work plane, ask yourself where this work plane needs to exist and what you know about this location. For example, you might want a plane to be tangent to a given face and parallel to another plane, or to go through the center of two arcs. Once you know what you want, select from the appropriate options and create a work plane. There are times when you may need to create an intermediate (construction) work plane before creating the final work plane. For example, you need to create a work plane that is at 30° and tangent to a cylindrical face. First, create a work plane that is at a 30° angle and located at the center of the cylinder, and then create a work plane parallel to the angled work plane that is also tangent to the cylinder.

To create a work plane, pick the Work Plane tool from the Feature Panel Bar as shown in Figure 3.56. Then, depending upon what, where, and how you select, a work plane will be created. Work planes can also be based on the default reference planes that exist in every part. The default planes have their visibility initially turned off, but can be turned on by expanding the Origin folder in the browser and then right clicking on a plane(s) and selecting Visibility from the context menu. These default planes can be used to create a New Sketch or used to create other work planes. Following is a list of methods that can be used to create a work plane.

Figure 3.56

- Select three points.
- Select a plane and a point.
- Select a plane and an edge or axis.
- Select two edges or two axes or an edge and axis.
- Create a work plane that is tangent to a face. Select a plane and face and the resulting work plane will be parallel to the selected plane and tangent to the selected face.
- To create an angled plane, select a plane and an edge; a dialog box will appear for you to type in the angle.
- To create a work plane that is offset a distance from another plane, select a plane and then drag the new work plane to a selected location. While you are dragging the work plane, an Offset dialog box will appear that will show the offset distance. Select a point, then type in a value for the offset distance and pick the check mark in the dialog box or press the ENTER key on the keyboard.

Note: The order in which points or planes are selected is irrelevant.

When creating a work plane and more than one solution is possible, a selection box will appear. Pick the forward or reverse arrows in the selection box until the correct solution is displayed. Then select the check mark in the selection box.

If a midpoint on an edge is selected, the resulting work plane will be linked to that midpoint. If the selected edge's length changes that the midpoint was selected from, the work plane's location will adjust to the new midpoint.

FEATURE VISIBILITY

The visibility of the origin planes, origin axes, origin point or user work planes, user work axes, user work points and sketches can be controlled by either right clicking on them in the graphics area or on its name in the Browser and then right click and pick Visibility from the context menu. The visibility can also be controlled for *all* origin planes, origin axes, origin point or user work planes, user work axes, user work points and sketches. From the View pull down menu pick Work Geometry and then check to turn the visibility on or uncheck to turn off the visibility of the selected set like what is shown in Figure 3.57.

The best way to get an understanding of work planes is to create them. In the tutorials that follow, you will create the most common types of work planes. Before each tutorial, there is a description of the type of work plane that will be created.

Figure 3.57

TUTORIAL 3.8—CREATING WORK PLANES

In this exercise you will create three work planes: one that is parallel to a plane and goes through a midpoint, one that goes through three points, and one that goes through two edges.

1. Start a new part file based on the default Standard.ipt file.
2. Sketch and dimension a rectangle that measures 2 inches horizontally by 1 inch vertically.
3. Change to an isometric view by selecting the right mouse button while in the graphics area and then select Isometric View from the context menu.
4. Extrude the sketch 3 inches.

5. Pick the Work Plane tool from the Features Panel Bar, select the front right face, and then select the midpoint of the top left edge as shown in Figure 3.58. The work plane will be created through the middle of the part.

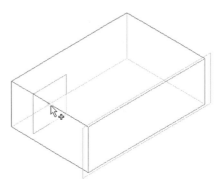

Figure 3.58

6. Create a New Sketch on the work plane you created.

7. Sketch and dimension a circle as shown in Figure 3.59. When placing the horizontal and vertical dimensions, select the circle and the outside edge of the part and then place the dimension. When the edge of the part is selected, it will be projected onto the active sketch.

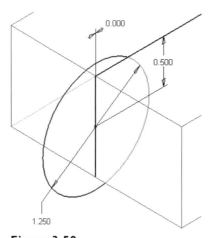

Figure 3.59

8. Extrude the circle .25 inch, removing material from the part using the midplane option.

9. Turn off the visibility of the work plane by either right clicking on the work plane in the graphics area or on its name in the Browser. Pick Visibility from the context menu, and when you are done, your part should resemble Figure 3.60.

Placed Features 125

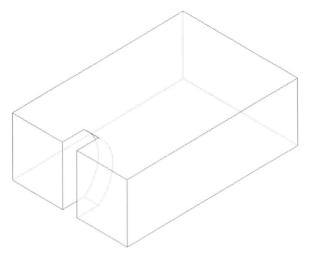

Figure 3.60

10. Edit the first sketch and change the 2-inch dimension to 3 inches. Update the part and you will see that the circular extrusion is still in the middle of the part.

11. Pick the Work Plane tool and select three points. Select two diagonal points from the top plane and a point on the front most edge on the bottom plane as shown in Figure 3.61.

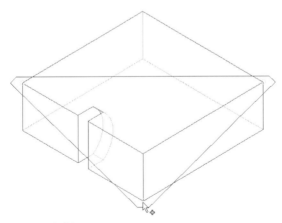

Figure 3.61

12. Turn off the visibility of the work plane by either right clicking on the work plane in the graphics area or on its name in the Browser and pick Visibility from the context menu.

13. Pick the Work Plane tool and select the two vertical edges as shown in Figure 3.62. For clarity, the second work plane's visibility was turned off.

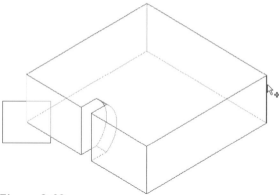

Figure 3.62

14. Practice making a New Sketch on each work plane and create a sketched feature. Edit the dimensions of the first extrusion and watch how the work planes maintain their relationship to the points that were used to create them.

15. When you are done practicing, close the file without saving the changes.

TUTORIAL 3.9—CREATING WORK PLANES

In this exercise you will create two work planes: one that goes through a work axis and is at an angle and another that is parallel to the first work plane and tangent to a circular face.

1. Start a new part file based on the default Standard.ipt file.

2. Sketch, dimension, and extrude the part as shown in Figure 3.63.

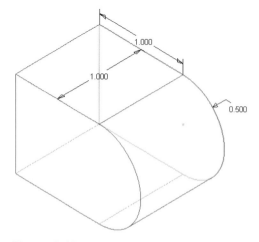

Figure 3.63

3. If you are not already in an isometric view, change to an isometric view by selecting the right mouse button while in the graphics area and then select Isometric View from the context menu.
4. Place a work axis through the center of the cylindrical face.
5. Pick the Work Plane tool and select the back vertical face and the work axis as shown in Figure 3.64. In the Angle dialog box, type in a value of 45.

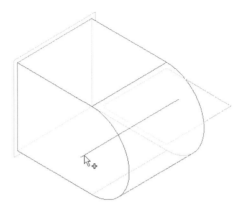

Figure 3.64

6. Pick the Work Plane tool and select the circular face and the work plane that you just created. A work plane will be created that is tangent to the circular face and parallel to the first work plane. Dynamically rotate the part to ensure that the second work plane is tangent to the circular face.
7. Make a New Sketch on the tangent work plane.
8. To make it easier to see the geometry, turn off the visibility of the work planes by picking the View pull down menu, then select Work Geometry and uncheck the User Work Plane set.
9. Sketch a rectangle similar to what is shown in Figure 3.65.

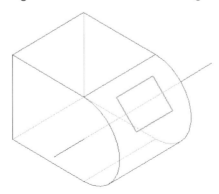

Figure 3.65

10. Extrude the rectangle .5 inch into the part, removing material.
11. Edit the first work plane by right clicking on the Work Plane1 feature in the browser and selecting Redefine Feature from the context menu. Reselect the back face of the part and the work axis. Change the angle to 15 degrees. If needed, update the part, and when you finish your part should look similar to Figure 3.66.

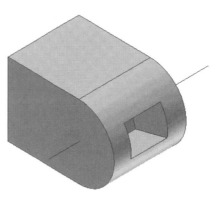

Figure 3.66

12. Edit the dimensions of the first extrusion and watch how the work planes maintain their relationship to the points that were used to create them.
13. When you are done practicing, close the file without saving the changes.

TUTORIAL 3.10—CREATING WORK PLANES

In this exercise you will create a work plane that is parallel to a reference plane and is tangent to a circular face. You will then place a tapped hole through the part.

1. Start a new part file based on the default Standard.ipt file.
2. Change to an isometric view by selecting the right mouse button while in the graphics area and then select Isometric View from the context menu.
3. Sketch and dimension a 1-inch diameter circle and extrude it 1 inch.
4. Expand the Origin folder in the browser.
5. Pick the Work Plane tool and select the circular face. In the browser select the XZ plane as shown in Figure 3.67.

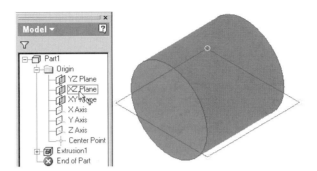

Figure 3.67

6. Make a New Sketch on the work plane. To make it easier to see the geometry, turn off the visibility of the work plane.

7. Place a Hole Center near the middle of the part and add the .5 inch vertical dimension as show in Figure 3.68. To center the hole horizontally in the cylinder, you will need to project objects. Projecting objects will be covered in chapter 6.

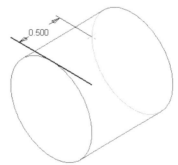

Figure 3.68

8. Use the Hole tool to create a 1/4-20 tapped hole that goes all the way through the part and has full threads. When done your part should resemble Figure 3.69.

Figure 3.69

9. Edit the dimensions of the first extrusion and watch how the work plane maintains its relationship to the circular face.

10. When you are done practicing, close the file without saving the changes.

PATTERNS

Once a feature is created and you need to duplicate the feature multiple times with a set distance or angle between them, you can use the pattern tool. There are two types of patterns, rectangular and circular. The pattern will be held together as a single feature in the browser, but the individual occurrences will be listed under the pattern feature. The entire pattern or individual occurrences can be suppressed. Both rectangular and circular patterns have a child relationship to the parent feature that was patterned. If the size of the parent feature changes, all of the child features will also change. If a hole is patterned, and the parent hole type changes, the child holes will also change. Since a pattern is a feature, it can be edited like any other feature. The base part or feature as well as patterns can be patterned. A rectangular pattern, will repeat the selected feature(s) along the direction set by two edges on the part (these edges do not need to be horizontal or vertical) and a polar pattern will repeat the feature(s) around an axis or edge. When creating a rectangular pattern, you will define two directions by selecting an edge that defines each alignment. Before creating a circular pattern, you must have a work axis, a part edge or a circular face in which the feature will be rotated about. After selecting the Rectangular or Circular tool from the Feature Panel Bar as shown in Figure 3.70 and 3.71, a pattern dialog box will appear, first select a feature to pattern and then type in the values and select the edges and axis as needed. If you move the dialog box away from the part, you will see a preview image of how the pattern will look. Below is a description of the options available for both rectangular and circular patterns.

Figure 3.70 — Rectangular Pattern

Figure 3.71 — Circular Pattern

RECTANGULAR PATTERNS

The rectangular pattern dialog box, shown in Figure 3.72, will be explained in this section.

Features: Pick this button and then select a feature or features to be patterned. Features can be removed from the selection set by holding down the CTRL key and selecting them.

Direction 1: Here you will define the first direction for the alignment of the pattern; it can be horizontal, vertical or angled.

> **Direction:** Pick this button and then select an edge that defines the alignment that the feature will be patterned along.

Flip: If the preview image shows the pattern going in the wrong direction, pick this button to reverse its direction.

Count: Type in a value or select the arrow to choose a previously used value that represents the number of times the feature(s) will be duplicated.

Spacing: Type in a value or select the arrow to choose a previously used value that represents the distance between the patterned features.

Direction 2: Here you will define the second direction for the alignment of the pattern; it can be horizontal, vertical, or angled but cannot be parallel to Direction 1.

Direction: Pick this button and then select an edge that defines the alignment that the feature will be patterned along.

Flip: If the preview image shows the pattern going in the wrong direction, pick this button to reverse its direction.

Count: Type in a value or select the arrow to choose a previously used value that represents the number of times the feature(s) will be duplicated.

Spacing: Type in a value or select the arrow to choose a previously used value that represents the distance between the patterned features.

Creation Method

Identical: When Identical is selected, all the occurrences in the pattern will use the same termination as the parent feature(s). This is the default option.

Adjust to Model: When Adjust to Model is selected, the termination of each occurrence will be calculated. Since each occurrence is calculated separately the processing time can be increased. This option must be used if the parent feature(s) terminates to a face or plane.

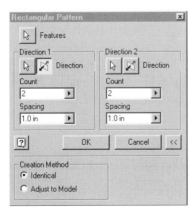

Figure 3.72

CIRCULAR PATTERNS

The circular pattern dialog box shown in Figure 3.73 will be explained in this section.

Features: Pick this button and then select a feature or features to be patterned. Features can be removed from the selection set by holding down the CTRL key and selecting them.

Rotation Axis: Pick the button and then select an edge, axis, or circular face (center) that defines the axis that the feature(s) will be rotated about.

Placement: Pick this button and then select an edge that defines the alignment that the feature will be patterned along.

Count: Type in a value or select the arrow to choose a previously used value that represents the number of times the feature(s) will be duplicated.

Angle: Type in a value or select the arrow to choose a previously used value that represents the angle that will be used to calculate the spacing of the patterned features.

Flip: If the preview image shows the pattern going in the wrong direction, pick this button to reverse its direction.

Creation Method
Identical: When Identical is selected, all the occurrences in the pattern will use the same termination as the parent feature(s). This is the default option.

Adjust to Model: When Adjust to Model is selected, each occurrence's termination will be calculated. Since each occurrence is calculated separately, the processing time can be increased. This option must be used if the parent feature(s) terminates to a face or plane.

Figure 3.73

Positioning Method
Incremental: When Incremental is selected, each occurrence will be separated by the number of degrees specified in the Angle area of the dialog box.

Fitted: When Fitted is selected, each occurrence will be evenly spaced within the angle specified in the Angle area of the dialog box.

 Note: A base part or feature can be patterned. Before creating a circular pattern, a work axis, edge, or a cylindrical face that the feature will be rotated about must exist.

TUTORIAL 3.11—RECTANGULAR PATTERNS

1. Open the file \Inventor Book\Rectangular Pattern.ipt
2. Pick the Rectangular Pattern tool from the Features Panel Bar and then select the blue cylinder and the red fillet as the features to be patterned.
3. Pick the Direction 1 button and select the bottom horizontal edge. If the direction is pointing to the left, reverse its direction by selecting the Flip button.
4. Change the Count to 4 and Spacing to .5. Then your screen should resemble Figure 3.74.

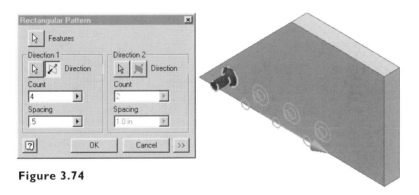

Figure 3.74

5. Select the Direction 2 button and select the left angled edge. If the direction is pointing down, reverse its direction by selecting the Flip button.
6. Leave the Count at 2 and change the Spacing to .75. Your screen should resemble Figure 3.75.

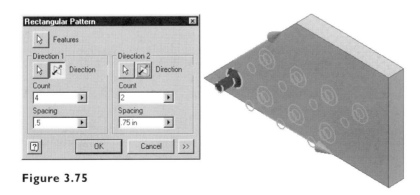

Figure 3.75

7. Select the OK button to create the pattern.

8. Edit the Extrusion distance of the blue cylinder (Extrusion2) and change from .25 inches to .5 inches and when done editing the dimension, update the part if needed. When done your screen should resemble Figure 3.76.

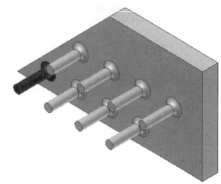

Figure 3.76

9. Practice editing the Pattern feature by trying different values for the count and spacing. Also edit the parent features and try different values and then delete the feature Fillet1.

10. When you are done practicing, close the file without saving the changes.

TUTORIAL 3.12—CIRCULAR PATTERNS

1. Open the file \Inventor Book\Work Axis and Work Point.ipt; this is the file that you added work axes to earlier in this chapter.

2. Pick the Circular Pattern tool and the select the hole that is on the top face as the feature to be patterned.

3. Pick the Rotation Axis button and select either the large circular face or the work axis that is in the center of the circular face as the axis of rotation.

4. Change the Count to 5 and leave the Angle at 360 degrees. Your screen should resemble Figure 3.77. Leave the other settings as shown.

 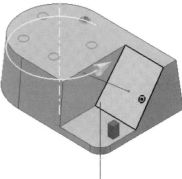

Figure 3.77

5. Select OK to create the pattern.
6. Pick the Circular Pattern tool and select the hole that is on the angled face as the feature to be patterned.
7. Pick the Rotation Axis button and select the work axis that is in the center of the angled face as the axis of rotation.
8. Change the Count to 5 and the Angle to 180 degrees and flip the direction. Your screen should resemble Figure 3.78. Leave the other settings as shown.

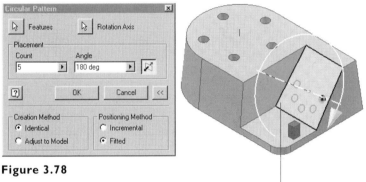

Figure 3.78

9. Select OK to create the pattern.
10. Pick the Circular Pattern tool and select the green rectangular extrusion as the feature to be patterned.
11. Pick the Rotation Axis button and select the work axis that is in the front of the part to use as the axis of rotation.
12. Change the Count to 3 and the Angle to 25 degrees. Change the Positioning Method to Incremental. Your screen should resemble Figure 3.79. Select OK to create the pattern.

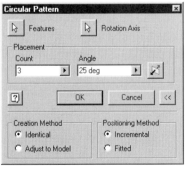
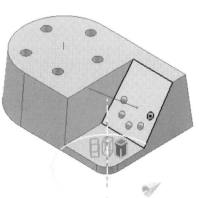

Figure 3.79

13. Practice editing the Pattern features by trying different values for the counts, angles and positioning methods.

14. When you are done practicing, close the file without saving the changes.

PRACTICE EXERCISES

The following exercises are intended to challenge you by providing problems that are open-ended, as the kinds of problems you'll encounter in work situations will be. There are multiple ways to arrive at the intended solution. In the end, your solution should match the sketch that is shown. Follow the directions for each of the exercises. When you have completed the part, save the file.

Exercise 3.1—Drain Plate

Start a new part file based on the default Standard.ipt file. Then create the part as detailed in Figure 3.80. In this exercise, utilize the shell, face draft, hole, and rectangular pattern tools that you learned about in this chapter. When done creating the part, save the file as \Inventor Book\Drain Plate.ipt. In chapter 4 you will create drawing views from the completed part.

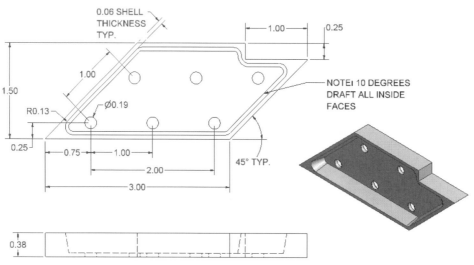

Figure 3.80

Exercise 3.2—Cylinder

Start a new part file based on the default Standard.ipt file. Then create the part as detailed in Figure 3.81. In this exercise utilize the work plane, hole, thread, chamfer, and circular pattern tools that you learned about in this chapter. When done, save the file as \Inventor Book\Connector.ipt. In chapter 4 you will create drawing views from the completed part.

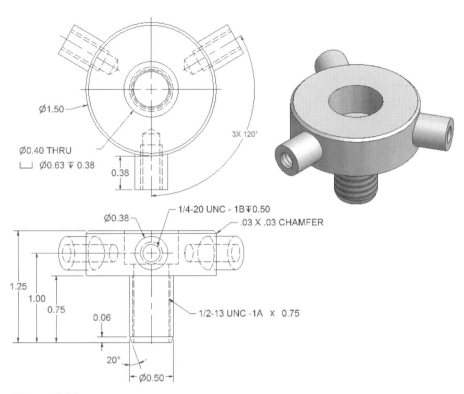

Figure 3.81

REVIEW QUESTIONS

1. When creating a fillet feature that has more than one selection set, each selection set will appear as an individual feature in the browser. True or False?

2. In regards to creating a fillet feature, what is a smooth radius transition?

3. When creating a fillet feature with the All Fillets option, material will be removed from all concave edges. True or False?

4. When creating a chamfer feature with the Distance and Angle option, only one edge can be chamfered at a time. True or False?

5. When creating a Hole feature, you do *not* need to have an active sketch. True or False?

6. What is a Hole Center used for?

7. Thread features are graphically represented on the part and will be annotated correctly when drawing views are generated. True or False?

8. A part may contain only one shell feature. True or False?

9. When creating a face draft feature, what does Pull Direction mean?

10. A work axis can only be created by selecting a cylindrical face. True or False?

11. Every new sketch needs to be derived from a work plane feature. True or False?

12. Explain how to create an offset work plane.

13. Work planes cannot be based from reference planes. True or False?

14. When creating a rectangular pattern, the directions that the feature will be duplicated along must be horizontal or vertical. True or False?

15. When creating a circular pattern, only a work axis can be used as the axis of rotation. True or False?

CHAPTER 4

Creating and Editing Drawing Views

After creating a part or assembly, the next step is to create 2-D drawing views that represent the part or assembly. To create drawing views, you will start a new a drawing file, select a 3-D part or an assembly to base the drawing views on, insert or create a drawing sheet with a border and title block, project orthographic views from the part or assembly, and then add annotations to the views. Drawing views can be created at any point after a part exists. The part does not need to be complete since the part and drawing views are associative in both directions (bidirectional). This means that if the part changes, the drawing views will automatically be updated when you return to the drawing views. If a parametric dimension changes in a drawing view, the part will get updated before the drawing views get updated. This chapter will guide you through these steps for setting up drawing standards and creating: borders, title blocks, drawing views of a single part, cleaning up the dimensions, adding annotations, and creating a bill of materials to complete the drawing. Chapter 5 will guide you through the steps for creating drawing views of assemblies and adding balloons.

CHAPTER OBJECTIVES

After completing this chapter, you will be able to:

- understand drawing options.
- create and edit drawing borders and title blocks.
- set up drawing standards.
- create drawing views from a part.
- edit the properties and location of drawing views.
- edit, move and hide dimensions.
- add reference dimensions.

- add annotations such as GD&T, surface finish symbols, weld symbols, datums and centerlines.
- create a bill of materials.
- create an AutoCAD 2-D Drawing from Inventor drawing views.

DRAWING OPTIONS, CREATING A DRAWING, AND DRAWING TOOLS

In this first section you will learn about the drawing options that are available, learn how to create a new drawing, and learn about the tools that will be used to create drawing views.

DRAWING OPTIONS

Before drawing views are created, the drawing options should be set to your preferences. To set the drawing options, pick Application Options from the Tools pull down menu. The Options dialog box will appear. Pick the Drawing tab and then your screen should resemble Figure 4.1. Make any changes to the options before creating the drawing views, otherwise the changes may not affect drawing views that have already been created. Following is a description of the drawing options.

Precise View Generation: If checked, drawing views will be displayed with their full precision when they are placed. Inventor allows you to continue working while it processes all of the entities in the view. If the check box is cleared, drawing views will be placed with an approximation of the geometry. This option requires less processing power and can allow you to work faster.

Get Model Dimensions on View Placement: When checked, applicable model dimensions are added to drawing views when they are placed. If the box is cleared, no model dimensions will automatically be placed. You can override this by manually selecting "Dimensions" in the "Create View" dialog box.

Show Line Weights: When checked, visible lines in drawings will be displayed with the line weights defined in the active drafting standard. If the check box is cleared, all visible lines are displayed with the same weight.

Note: This setting does not affect line weights when the views are printed.

Save DWF Format: When checked, DWF information is saved in the IDW file. The DWF file can be viewed by a web browser using the WHIP plug-in from Autodesk. If the check box is cleared, no DWF information will be saved in the IDW file. Saving DWF information in the file will slightly increase the size of the file.

Note: For information about the Whip! viewer, visit the Autodesk Web site at www.autodesk.com

Highlight Invalid Annotations: When checked, annotations that no longer are valid for the drawing view(s) will be highlighted.

Alternate Title Block Alignment: Pick the location where the title block will be placed.

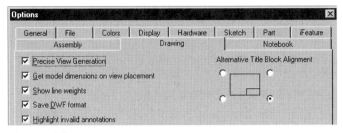

Figure 4.1

CREATING A DRAWING

The first step in creating a drawing from an existing part or assembly is to create a new drawing IDW file by either selecting the New icon from the What To Do section and then select the Standard.idw icon from the Default template tab as shown in Figure 4.2 or select New from the File pull down menu or from the left side of the Standard toolbar select the down arrow on the New icon and select Drawing.

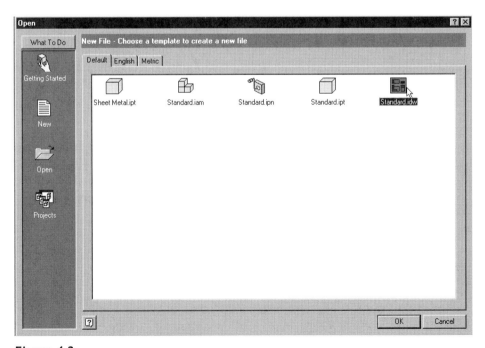

Figure 4.2

DRAWING TOOLS

After starting a new drawing, Inventor's screen will change to reflect the new drawing environment. There are three toolbars available in the Panel Bar that you will use to create drawings: Drawing Management, Sketch, and Drawing Annotation. By default when a new drawing is created, the Panel Bar will show the Drawing Management toolbar. These tools will allow you to create drawing views and sheets. The Sketch tools will enable you to add geometry and dimensions to draft views. The Drawing Annotation tools enable you to add annotations to existing drawing views. The three sets of tools will be explained throughout this chapter.

DRAWING SHEETS PREPARATION

When you start a drawing file, it displays a default drawing sheet with the default title block and border. The default drawing sheet, title block, and border are set from the template IDW file that was selected. The drawing sheet represents a blank piece of paper on which the border, title block, and drawing views will be placed. There is no limit to the number of sheets that can exist in the same drawing, but you must have at least one drawing sheet. To create a new sheet, pick the new Sheet tool from the Drawing Management Panel Bar as shown in Figure 4.3 or right click in the Browser and pick new Sheet from the context menu. A new blank sheet will appear in the Browser and the new sheet will be shown in the graphics area without a border or title block. In the next section you will learn how to add a border and title block. The sheet can be renamed by slowly double clicking on its name in the Browser and then typing in a new name or right click on the sheet in the Browser and pick Edit Sheet from the context menu. Then type in a new name in the Edit Sheet dialog box. To create a sheet of a different size you can select a Sheet Format from the Drawing Resources area in the Browser as shown in Figure 4.4. If a sheet is selected from the Sheet Formats list, predetermined drawing views will be created. If the sheet size that you need is not on the list, right click on the sheet name in the Browser and pick Edit Sheet from the context menu. Then in the Edit Sheet dialog box, click in the Size area and select a size from the drop list as shown in Figure 4.5. To use your own values, click Custom Size from the list and then enter values for the Height and Width.

Figure 4.3

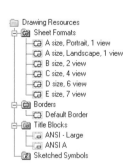

Figure 4.4

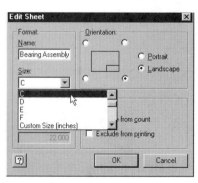

Figure 4.5

Note: The sheet size is inserted full scale (1=1) and should be plotted at 1=1. The drawing views will be scaled to fit the sheet size.

BORDER CREATION

The four lines and zone labels that surround the edges of the sheet make up the border. To insert the default border to a blank drawing sheet, expand the Drawing Resources name in the Browser and double click on Default Border. To insert a customized border with specified margins, expand the Drawing Resources name in the Browser and right click on Default Border and select Insert Drawing Border. The Default Drawing Border Parameter dialog box will appear like what is shown in Figure 4.6. Select the options and type in the values that you want and then select the OK button. The border will then appear on the sheet. The Sheet Margins will be offset from the edges of the sheet. To delete an existing border, in the Browser expand the sheet name, right click on the Border's name, and pick Delete from the menu.

Figure 4.6

TITLE BLOCK CREATION

To add a title block to the drawing sheet, you can either insert a default title block or you can construct a customized title block and insert it onto the drawing sheet.

DEFAULT TITLE BLOCK

To insert a default title block follow these guidelines.

- Make the sheet active, then place the title block by double clicking on its name in the Browser.

- Insert the title block by expanding Drawing Resources>Title Blocks in the Browser and then either double click on the title block's name as shown in Figure 4.7 or right click on the title block's name and select Insert from the context menu.

The title block will be inserted on the current sheet. By default the information in the title block will automatically be filled in from the properties of the drawing file. The file properties for the drawing can be filled in by either right clicking on the top level drawing name in the Browser and pick Properties or from the file pull down menu pick Properties. The Drawing Properties dialog box will appear and you can fill in the information as needed. Figure 4.8 shows the Drawing Properties dialog box and the result in the title block. Chapter 7 and 8 will cover other uses for file Properties. Note, the fields in the title block can be customized to get properties from the part or assembly file by adjusting the Type.

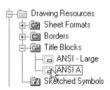

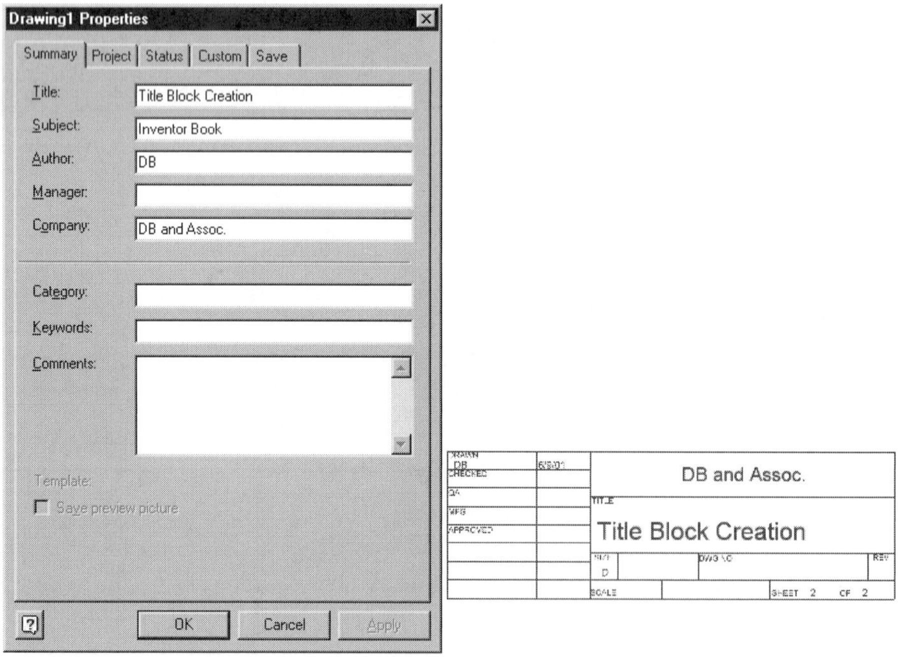

Figure 4.7

Figure 4.8

CREATING A NEW TITLE BLOCK

In this section you will be introduced to the methods of creating a customized title block. The book however will utilize the default title blocks. If you want to create a customized title block you can either edit an existing title block or create one from scratch. To edit an existing title block, right click on the title block name in the Browser under Drawing Resources>Title Block and pick Edit from the menu. The title block will appear with all the attributes shown. Edit as needed and when complete, right click and pick Save Title Block and then you will have the option to save it as another name. If the title block is saved as another name, the new title blocks name will appear in the Browser under Drawing Resources>Title Block. The new title block will not be available to new drawings unless it is saved in a template file.

If you want to create a new title block from scratch right click on Title Blocks in the Browser under Drawing Resources and select Define New Title Block. Next, using the tools from the Sketch toolbar, create the title block. Create the title block using the sketch tools in the same fashion as you would when creating a sketch. To create text that is populated from the file properties, use the Property Field tool. When done creating the title block, right click and pick Save Title Block from the menu. The title block will appear in the Browser under Drawing Resources>Title Block. The new title block will not be available to new drawings unless it is saved in a template file. See the online help to learn how to customize a title block's properties.

TUTORIAL 4.1—SHEETS, BORDERS, AND TITLE BLOCKS

In this tutorial you will start a new drawing, edit the sheet size, insert a customized border, and place an ANSI title block. The drawing will be used in the tutorials for this chapter.

1. Start a new drawing file based on the default Standard.idw file.
2. Edit the sheet name and size by right clicking on Sheet:1 in the Browser and pick Edit Sheet from the menu. In the Edit Sheet dialog box change the Name to "A Size" and pick "A" from the Size drop list like what is shown in Figure 4.9. Then Click OK to close the Edit Sheet dialog box.

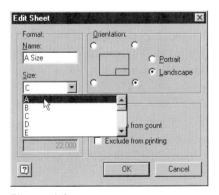

Figure 4.9

3. Delete the existing border by right clicking in the Browser on the name "Default Border" under the sheet name "A Size:1" and then pick Delete from the menu.

4. Delete the existing title block by right clicking in the Browser on the name "ANSI Large" under the sheet name "A Size:1" and then pick Delete from the menu.

5. Insert a border and edit the margins by expanding the Drawing Resources and Borders Folders in the Browser, then right click on Default Border and select Insert Drawing Border. In the Default Drawing Border Parameters dialog box change the Horizontal and Vertical Zone to 0, uncheck Center marks from the More section of the dialog and set the sheet margins to .4 for the Top, Bottom, and Right margins, then set the Left margin to .6 as shown in Figure 4.10. Click OK to place the border on the drawing sheet.

Figure 4.10

6. To populate the title block, change the file Properties of the drawing by either right clicking on the top level drawing name in the Browser and pick Properties from the menu or from the File pull down menu pick Properties. In the Drawing Properties dialog box, fill in your own information for Title, Author, and Company name. Figure 4.11 shows an example of the filled in Drawing Properties dialog box.

7. Insert the ANSI A title block by double clicking on ANSI A in the Browser under Drawing Resources>Title Blocks. When complete your screen should resemble Figure 4.12.

8. Save the file as "\Inventor Book\Chapter 4 Drawings.idw." This file will be used in the next tutorial.

Creating and Editing Drawing Views   147

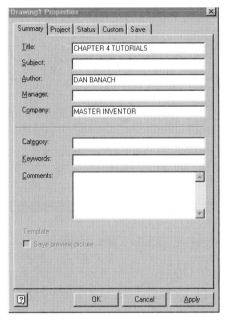

Figure 4.11

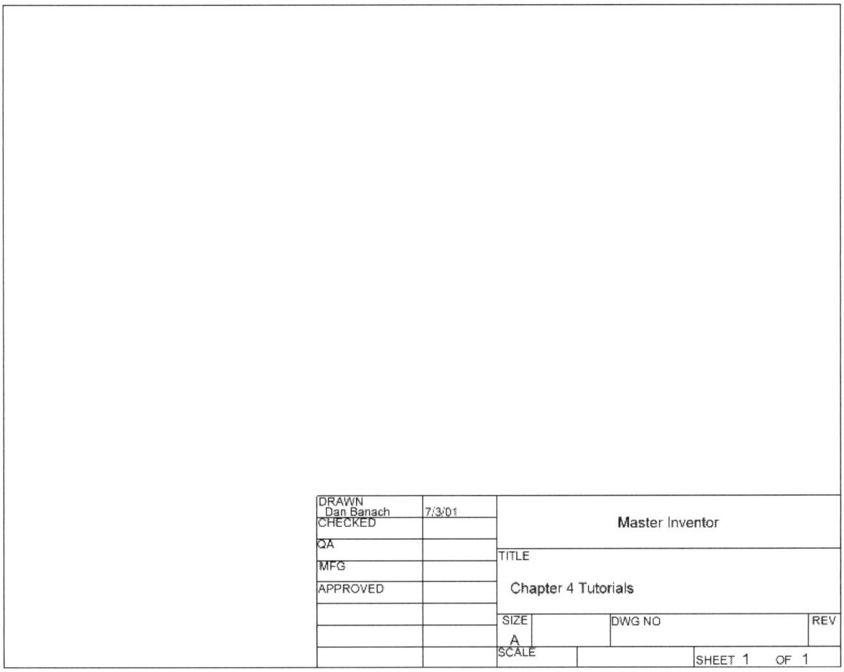

Figure 4.12

SETTING UP DRAFTING AND DIMENSION STANDARDS
DRAFTING STANDARDS
Before creating drawing views, drafting standards should be set or configured. There are six drafting standards that are already configured and included with Inventor: ANSI, BSI, DIN, GB, ISO, and JIS. You can choose to use one of these standards or create your own standard based on one of them. To make a standard current, pick Standards from the Format pull down menu. The Drafting Standards dialog box will appear like what is shown in Figure 4.13. From the Select Standard list, pick the standard that you want to use.

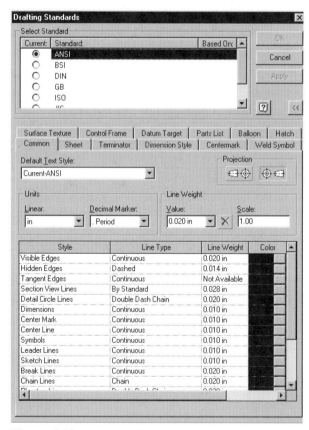

Figure 4.13

New Drafting Standard
The default drafting standards cannot be modified. If you need to modify a drafting standard, you will need to create a new drafting standard based on one of the included standards. If a new drafting standard is created, the changes will apply only

to that drawing. To make the new drafting standard available for other new drawing files, make the changes in the template that you use to create new drawings and then save the file. To automatically have the new drafting standard be the default standard, set the new standard as the active standard.

To create a new drafting standard follow these guidelines.

- From the Format pull down menu pick Standards.
- In the Select Standard area of the Drafting Standards dialog box, pick Click to add new standard at the bottom of the Standard list.
- In the New Standard dialog box type in a name for the new standard and select the Standard that the new style will be based on.
- In the Drafting Standards dialog box pick the >> button to expand the dialog, if necessary, and twelve tabs will appear. Make the changes as needed and when done pick the Apply button.

Following is a description of the tabs for the Drafting Standards dialog box.

- **Common:** Controls the text style, projection direction, units of measurement, and line style
- **Sheet:** Controls the sheet and view labels in the browser bar and the color scheme of the drawing sheet
- **Terminator:** Controls the style of arrows and datum references
- **Dimension Style:** Controls the default dimension style and defines which characters will be available to use in dimension text
- **Centermark:** Controls the proportion of the dashes used in a centermark
- **Weld Symbol:** Controls the weld symbols
- **Surface Texture:** Controls the surface texture symbols
- **Control Frame:** Controls the control frame of geometric tolerances
- **Datum Target:** Controls the datum target of geometric tolerances
- **Parts list:** Controls the styles of the parts list
- **Balloon:** Controls the styles of the balloons
- **Hatch:** Controls the styles of the hatch patterns

DIMENSION STYLES

To control how dimensions will be displayed when a drawing view is created, you set the properties of a dimension style. There are six dimension styles (based on the six drafting standards) that are already configured and included with Inventor: ANSI, BSI, DIN, GB, ISO, and JIS. You can choose to use one of these dimension standards or create your own standard based on one of them. To make a dimension style current, pick Dimension Style from the Format pull down menu. The Dimension Styles dialog box will appear as shown in Figure 4.14. From the Standard drop list pick the standard that you want to use. Depending upon the standard that you choose, there may be a style to choose from under the Style Name area. Example would be ANSI Fraction or Default-ANSI (two place decimal).

 Note: The book will use the Default-ANSI dimension style.

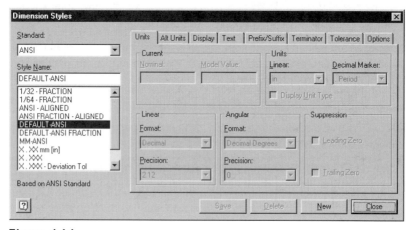

Figure 4.14

New Dimension Style

The default dimension styles cannot be modified. If you need to modify a dimension style you will need to create a new dimension style based on one of the included standards. After creating a new dimension style that will be the default style, set it as the active dimension style in the Drafting Standards dialog box under the Dimension Style tab. To create a new dimension style, follow these guidelines.

- From the Format pull down menu, pick Dimension Style.
- In the Dimension Styles dialog box, pick a standard that the new style will be based on.
- Pick the New button and type in a new name for the Style.
- Modify the style by changing the items under the tabs in the Dimension Style dialog box.
- When done modifying the new dimension style, pick the Save button.

Following is a description of the tabs for the Dimension Style dialog box.

- **Units:** Enables the units of measure that will be used for the dimensions in the drawing.
- **Alt Units:** Controls if and how alternate units will be displayed.
- **Display:** Controls the display characteristics for dimension lines in a drawing.
- **Text:** Controls the font, size, and format style for the default text.
- **Prefix/Suffix:** Controls the symbol, text, and location for a prefix/suffix before and after or above and below dimension text.
- **Terminator:** Controls the default arrowhead style for the drafting standard.
- **Tolerance:** Controls how a tolerance will be displayed.
- **Options:** Controls dimension style options. The options you select define the appearance of dimension lines and leaders in your drawing. The list of options available depends on the method and type you selected. Select as many options as desired from the list.

TEMPLATE

After you have created a drawing sheet, border, title block, and have set the drafting standard and dimension style, you can save the file as a template and start new drawing files based on these settings. To save a file as a template, pick Save Copy As from the File menu and save the file in the template directory of Inventor. Then use that file when creating new drawings.

CREATING DRAWING VIEWS

After the drawing sheet and standards have been set, drawing views are ready to be created from an existing part, assembly, or presentation file. The file that the views will be created from do not need to be opened when a drawing view is created from it. However, it is suggested that both the file and the associated drawing file be stored in the same directory and that the directory should be contained in the project file. When creating drawing views, you will find that there are many different types of drawing views that can be created.

- **Base View:** The first drawing view of an existing part, assembly or presentation file.
- **Projected View:** An orthographic or isometric view that is generated from an existing drawing view.

    Orthographic (Ortho View): A drawing view that is projected horizontally or vertically from another view.

    Iso View: A drawing view that is projected at a 30° angle from a given view. An isometric view can be projected to any of the four quadrants.

- **Auxiliary View:** A drawing view that is perpendicular to a selected edge of another view.
- **Section View:** A drawing view that represents the area defined by slicing a plane through a part or assembly.
- **Detail View:** A drawing view in which a selected area of an existing view will be generated at a specified scale.
- **Broken View:** A drawing view that shows a section of the part removed, but the ends remain. Any dimension that spans over the break will reflect the correct length.

In the following sections the different drawing views and how they can be created will be explained.

BASE VIEWS

A base view is the first view that is being created from the selected part, assembly, or presentation file. When you create a base view, the scale is set in the dialog box. From the base view, other drawing views can be projected. Projected drawing views will be explained in the next section. There is no limit to the number of base views that can be created in a drawing (based on different parts, assemblies, or presentation files. When creating a base view, the orientation of that view can be selected from the following list: Front, Current (current orientation of how the part, assembly, or presentation is viewed in its file), Top, Bottom, Left, Right, Back, Iso Top Right, Iso Top Left, Iso Bottom Right, and Iso Bottom left.

To create a drawing base view, follow these guidelines.

1. Pick the Create View tool from the Drawing Management Panel Bar as shown in Figure 4.15 or right click in the graphics area and pick Create View from the context menu. The Create View dialog box will appear as shown in Figure 4.16. From within the dialog box follow steps 2 to 5.

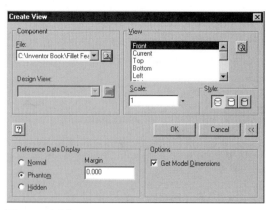

Figure 4.15

Figure 4.16

2. Pick the Explore directories icon to navigate to and select the part, assembly, or presentation file that the drawing views will be created or projected from. After making the selection, a preview image will appear in the graphics area, DO NOT place the view until all the options have been set.

3. Pick the type of view to generate from the View list.

4. Pick the scale for the view.

5. Pick the Style for the view.

6. Locate the view by picking a point in the graphics area.

The Create View dialog box is divided into two main areas; Component and View; each will be described in this section.

Components

In this section the component that the drawing views will be created from is selected.

File: Any open part, assembly, or presentation files will appear in the drop down list. Or pick the Explore directories icon and navigate to and select a part, assembly, or presentation file.

Design View / Presentation View: After selecting a file that has Design Views, the names of the Design Views will appear in the drop list. If a file has Presentation Views, the names of the Presentation Views will appear in the drop list

View

In this section, the type of view, scale, and style of the drawing view will be set.

View: After selecting a part, assembly, or presentation file you can choose the orientation that the view will be created. After picking an orientation, a preview image will appear in the graphics area. If the preview image does not show the view orientation that you wanted, select a different orientation view by picking another point. The following list shows the different view orientations that are available: Front, Current (current orientation of how the part, assembly, or presentation is viewed in its file), Top, Bottom, Left, Right, Back, Iso Top Right, Iso Top Left, Iso Bottom Right, and Iso Bottom left.

Scale: Type in a number for the scale that the view will be created. Note that the drawing will be plotted at full scale (1=1) and the drawing views are scaled as needed. The scale of the view(s) can be edited after the view(s) have been generated.

Style: In the Style area, choose how the view will be displayed. There are three choices: Hidden Line, Hidden Line Removed, and Shaded. The preview image will not represent the style choice. When the view is created, the chosen style will appear. The style can be edited after the view has been generated.

Reference Data Display

In this section the display of reference data will be set.

Normal: When checked, reference data will have no special display characteristics.

Phantom: When checked and in wireframe display, the reference data and product data will not hide each other. The reference data will be displayed on top of product data, and the reference data will be displayed with tangent edges turned off and all other edges as phantom linetype. When checked and in shaded display, the reference data will be displayed as wireframe with tangent edges turned off and all other edges as phantom. The reference data will be displayed on top of product data that will be shaded.

Hidden: When checked reference data will not be displayed.

Margin: To see more reference data, set the value to expand the boundaries by a specified value on all sides.

Options

In this section, the visibility of model dimensions for the view will be set.

Get Model Dimensions: Select the check box to see model dimensions in the view when it is created. If the box is unchecked, model dimensions will not automatically be placed upon view creation. When checked, only the dimensions that are planar to this view and have not been used in existing views on the sheet will display.

PROJECTED VIEWS

A projected view can be an orthographic or isometric view that is projected from a base view or any other existing view. When creating a projected view, a preview image will appear showing the orientation of view that will be created as the mouse is moved to a location on the drawing. There is no limit to the number of projected views that can be created. To create a projected drawing view, follow these guidelines.

- Pick the Projected View tool from the Drawing Management Panel Bar as shown in Figure 4.17, or right mouse click inside the bounding area of an existing view box (shown as dashed lines when the mouse is moved into the view) and pick Create View>Projected from the context menu.

Figure 4.17

- If the Projected View tool was picked from the Drawing Management toolbar, pick inside the view that the view will be projected from.

- Move the mouse horizontally, vertically, or at an angle to get a preview image of the view that will be generated. Keep moving the mouse until the review matches the view that you want to create then press the left mouse button. Continue placing projected views.

- When done placing views, right mouse click and pick Create from the context menu.

AUXILIARY VIEW

An auxiliary view is a view that is projected perpendicular from a selected edge or line. To create an auxiliary drawing view, follow these guidelines.

1. Pick the Auxiliary View tool from the Drawing Management Panel Bar as shown in Figure 4.18 or right mouse click inside the bounding area of an existing view box (shown as dashed lines when the mouse is moved into the view) and pick Create View>Auxiliary from the context menu.

    Figure 4.18

2. The Auxiliary View dialog box will appear; type in a name for the Label and a value for the Scale and select a Style.

3. If the Auxiliary View tool was picked from the Drawing Management toolbar, pick inside the view that the auxiliary view will be projected from.

4. In the selected drawing view, pick an edge or line that the auxiliary view will be projected perpendicular from and then move the mouse to position the view.

5. Click a point on the drawing sheet to create the auxiliary view.

SECTION VIEW

A section view is a view that is created by sketching a line or multiple lines that will define the plane(s) that will be cut through a part or assembly. The view will represent the surface area of the cut. When defining a section, you will sketch line segments that can be horizontal, vertical, or at an angle. Arcs, splines, and circles cannot be used to define section lines. When sketching the section line(s), geometric constraints will automatically be applied between the line being sketched and the geometry in the drawing view. You can also infer points by moving the mouse over certain geometry locations (such as centers of arcs, endpoints of lines, etc.), then moving the mouse away displays a dotted line showing that you are inferring or tracking that point. To place a geometric constraint between the drawing view geometry and the section line, click in the drawing when a green circle appears and you see the glyph for the constraint. If you do not want the section lines to automatically have constraint(s) applied to them when they are being created, hold down the CTRL key when sketching the line(s). Since the area in the section view that shows where there is solid material is displayed with a hatch pattern, the hatching style should be set before the section view is created. To create a section drawing view, follow these guidelines.

1. Pick the Section View tool from the Drawing Management toolbar as shown in Figure 4.19 or right mouse click inside the bounding area of an existing view box (shown as dashed lines when the mouse is moved into the view) and pick Create View>Section from the context menu.

    Figure 4.19

2. If the Section View tool was picked from the Drawing Management Panel Bar, pick inside the view that the section view will be created from.

3. Sketch the line or lines that define where and how you want the view to be cut.

4. When done sketching the section line, right click and pick Continue from the context menu.

5. The Section View dialog box will appear. Fill in the information for how you want the Label, Scale, and Style to appear in the drawing view and then pick the OK button to create the view.

6. After the view has been created, constraints and/or dimensions can be added or deleted to the section line(s) to better define the section cut. To add or delete a constraint or dimension, select the section line, right click and pick Edit from the context menu.

7. From the Sketch toolbar use the General Dimension and Constraint tools to add or delete constraints and dimensions.

8. When you are done editing the section line, right click and pick Finish Sketch or pick Return on the Command bar.

9. If the section lines were not constrained, they can be dragged to show the section in a new location. After dragging the section lines the dependent section view will be updated to reflect the change.

DETAIL VIEW

A detail view is a drawing view that enlarges an area of an existing drawing view by a specified amount. The enlarged area is defined by a circle (or rectangle) and can be placed anywhere on the sheet. To create a detail drawing view, follow these guidelines.

1. Pick the Detail View tool from the Drawing Management Panel Bar as shown in Figure 4.20 or right mouse click inside the bounding area of an existing view box (shown as dashed lines when the mouse is moved into the view) and pick Create View>Detail from the context menu.

Figure 4.20

2. If the Detail View tool was picked from the Drawing Management toolbar, click inside the view that the detail view will be created from.

3. The Detail View dialog box will appear; fill in information for how you want the Label, Scale, and Style to appear in the drawing view. *Do not* pick the OK button at this time as this will end the operation.

4. In the selected view, pick a point that will be used as the center of the circle that will describe the detail area.

Note: You can also right click at this point and choose Rectangular Fence to change the bounding area of the detail area definition.

5. Pick another point that will define the radius of the detail circle. As you move the mouse, a circle will appear that shows the size of the circle.

6. Pick a point on the sheet where you want the view to be placed. There are no restrictions on where the view can be placed.

BROKEN VIEW

When creating drawing views of long parts, you may want to remove a section or multiple sections from the middle of the part and show just the ends; this type of view is referred to as a broken view. For example, you may want to create a drawing view of a 2″ x 2″ x 1/4″ angle that is 5 feet long and only has the ends chamfered. When a drawing view would be created, the detail of the ends would be small and difficult to see since the part is so long. In this case you can create a broken view that removes a 4-foot section from the middle of the angle and leave 6 inches on each end. When an overall length dimension would be placed that spans over the break, it would be displayed as 5 feet and the dimension will have a break symbol to note that it is derived from a broken view. A broken view is created by altering an existing drawing view by adding breaks to it. The view types that can be changed into broken views are: part views, assembly views, projected views, isometric views, auxiliary views, section views, and detail views. After creating a broken view, the breaks can be moved dynamically to change what is seen in the broken view.

To create a broken drawing view, follow these guidelines.

1. First create a base view that will eventually be shown broken.

2. Pick the Broken View tool from the Drawing Management Panel Bar as shown in Figure 4.21 or right mouse click inside the bounding area of an existing view box (shown as dashed lines when the mouse is moved into the view) and pick Create View>Broken from the context menu.

3. If the Broken View tool was picked from the Drawing Management toolbar, pick inside the view that the broken view will be created from.

4. The Broken View dialog box will appear like what is shown in Figure 4.22. *Do not* pick the OK button at this time; this will end the operation.

5. Then in the drawing view that will be broken, pick a point where the break will begin and then pick a second point to locate the second break. As you move the mouse a preview image will appear showing the placement of the second break line.

6. To create another break repeat steps 2 through 5.

7. To edit the properties of the broken view, move the mouse over the break lines and a green circle will appear in the middle of the break. Right click and pick Edit Break from the context menu. The same Broken View dialog box will appear; edit the data as needed and when done pick the OK button to complete the edit.

8. To move the break lines, pick on one of the break lines and with the mouse button pressed down, drag the break line to a new location. The other break line will follow to maintain the gap size.

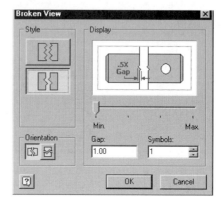

Figure 4.21

Figure 4.22

DRAFT VIEWS

Another type of drawing view is a draft view. A draft view is not created from a 3-D part, but contains one or more associated 2-D sketches. A draft view can be created on its own sheet, in its own IDW file or can be used in an existing sheet with other drawing views to provide detail that is missing in a model. From the Sketch Panel Bar you can add geometry, constraints, parametric dimensions, text, and other annotations in the draft view. All these objects will be placed in the draft view and if the draft view moves so will all the objects associated to it. When an AutoCAD 2-D file is imported into an Inventor drawing the geometry is placed in a draft view. To create a draft view, follow these guidelines.

1. Pick the Draft View tool from the Drawing Management toolbar as shown in Figure 4.23.

   Figure 4.23

2. The Draft View dialog box will appear, fill in information for how you want the Label and Scale of the Draft view to appear in the drawing view. Then pick the OK button to continue. The Scale that you set will be the scale that you are drafting the entities to.

3. No other dialog box or window will appear. Using the tools from the Sketch Panel Bar, create 2-D geometry and add constraints, dimensions and annotations as needed.

4. When done creating the geometry that should go into the draft view, either right click and pick Finish Sketch from the context menu or pick the Return button from the Command bar.

5. The draft view will then act like a regular drawing view and can be edited as such.

TUTORIAL 4.2—DRAWING VIEWS

In this tutorial you will create different types of drawing views.

1. If it is not already open, open the file \Inventor Book\Chapter 4 Drawings.idw

2. To create a base view, pick the Create View tool from the Drawing Management Panel Bar and pick the file \Inventor Book\Drawing View Part.ipt and then enter a scale of .5, select Get Model Dimensions from the More section of the dialog and then place the view as shown in Figure 4.24.

Note: If the dimensions did not appear as a two-place decimal, set Default-ANSI as the default dimension style from the Standards and recreate the views.

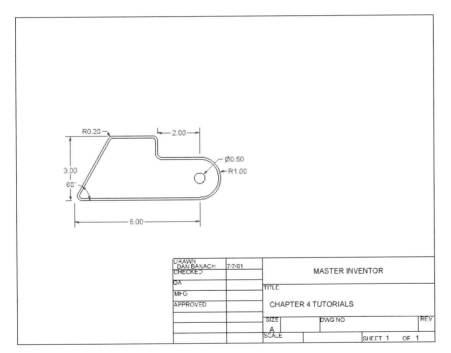

Figure 4.24

3. Project a top and isometric view using the Project view tool. When done, your drawing views should resemble Figure 4.25.

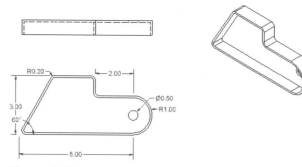

Figure 4.25

4. Create a section view of the front view by picking the Section View tool from the Drawing Management Panel Bar and pick a point in the front view. Then pick a point near the upper left side of the view and then pick a point below it to create a vertical line like what is shown in Figure 4.26. Then right mouse click and pick Continue from the context menu. Then locate the view by picking a point to right of the view. When done, your views should resemble Figure 4.27.

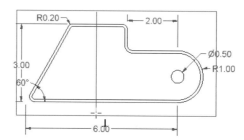

Figure 4.26

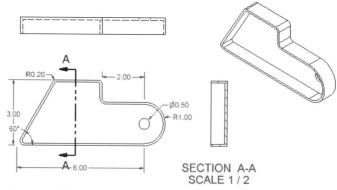

Figure 4.27

5. Create an isometric view of the section view in the upper right corner of the drawing sheet. When done, your views should resemble Figure 4.28.

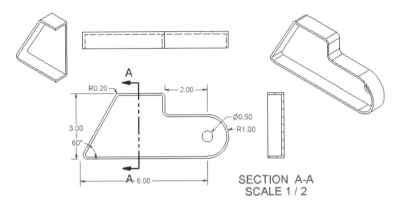

Figure 4.28

6. Move the section line in the front view by picking on it and dragging it to a new location. The section and isometric of the section should update to reflect the new location of the section. When done, move the section line back to its original position.

7. Create a detail view of the lower left corner of the front view by picking the Detail tool from the Drawing Management Panel Bar and pick a point in the front view. Then in the Detail View dialog box, change the scale to 2, then pick a point near the lower left corner of the view. That point will be used as the center of the detail. Then drag the circle so it resembles Figure 4.29 and then pick a point to define the details size. Place the detail in the lower left corner of the drawing, and when done your drawing should resemble Figure 4.30.

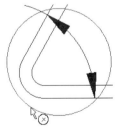

Figure 4.29

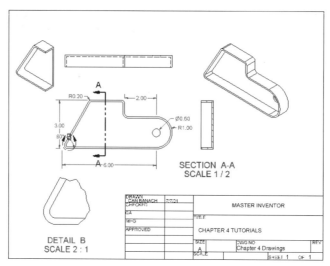

Figure 4.30

8. In the same file create a new sheet by picking the new Sheet tool from the Drawing Management Panel Bar or right click in the Browser and pick new Sheet from the context menu.

9. To create a base view, pick the Create View tool from the Drawing Management Panel Bar. Then in the Create View dialog box, pick the file \Inventor Book\Shaft.ipt and enter a scale of .5 and change the View type to Left, and select Get Model Dimensions from the More section of the dialog box. Then place the view as shown in Figure 4.31.

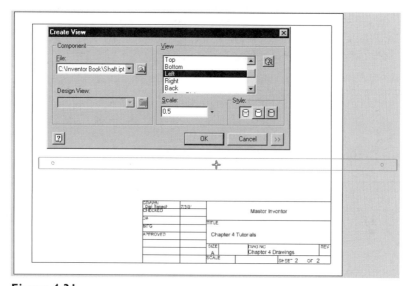

Figure 4.31

10. To alter the base view into a broken view, pick the Broken View tool from the Drawing Management Panel Bar and pick in the base view. Then using the defaults in the Broken View dialog box and in the drawing view, pick a point near the left end of the shaft but to the right of the hole, then pick a point to the left of the middle hole as shown in Figure 4.32. After picking the second point, the drawing view will then be broken in the picked locations.

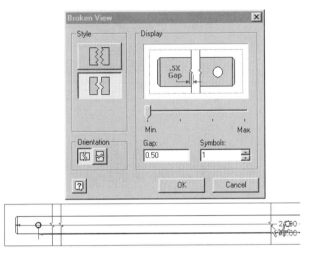

Figure 4.32

11. To break the right side of the drawing view, again pick the Broken View tool and pick in the base view. Then in the Broken View dialog box, change the Style to Rectangular and in the drawing view pick a point near the right end of the shaft but to the left of the hole, then pick a point to the right of the middle hole as shown in Figure 4.33. After picking the second point, the drawing view will then be broken in the picked locations. When done, your drawing view should resemble Figure 4.34.

 Note: The dimensions will be cleaned up in the next tutorial.

12. Alter the location of the break lines by dragging them to new locations. When done practicing, move the break lines so they resemble Figure 4.34.

13. When done, save the file; this file will be used in the next tutorial.

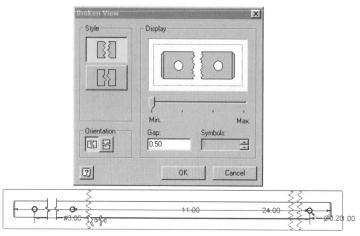

Figure 4.33

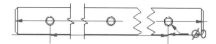

Figure 4.34

EDITING DRAWING VIEWS

After creating the drawing views, you may need to move, edit properties, or delete a drawing view. Each will be discussed in this section.

MOVING DRAWING VIEWS

To move a drawing view, position the mouse in that view until the bounding box appears. Then press and hold down the left mouse button and move the view to its new location, then release the mouse button. As the view is moved a rectangle will appear that represents the bounding box of the drawing view. If you move a base view any projected (children or dependent) views will also move with it. If an orthographic or auxiliary view is moved, it will only be able to be moved along the axis in which it was projected. Detail and isometric views can be moved anywhere in the drawing sheet.

EDIT DRAWING VIEW PROPERTIES

After creating a drawing view, you may need to change the label, scale, style, hatching visibility, remove the alignment constraint to its base view, or control the visibility of the view projection lines for a section or auxiliary view. To edit a drawing view, follow one of these guidelines.

- Double click in the bounding area of the view.
- Double click on the view's icon that you want to edit in the Browser.

- Right mouse click in the drawing view's bounding area or on its name and pick Edit View from the context menu.
- Right mouse click on the drawing view's name in the Browser and pick Edit View from the context menu.

Then the Edit View dialog box will appear similar to what is shown in Figure 4.35. Depending on the view that is selected, certain options may be grayed out from the dialog box. Make the needed changes and then pick the OK button to complete the edit. When the scale is changed in the base view all the dependent views will be scaled as well. Following is a description of the options that are available in the Edit View dialog box.

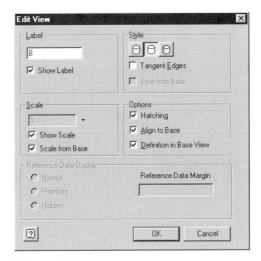

Figure 4.35

Label: Type in a label name and check the Show Label box if the label should be visible in the view.

Style: Pick the icon for the style that the view should be displayed in: Hidden Line, Hidden Line removed, or Shaded. Check the Tangent box if you want the tangent edges in the view to be displayed. If the Style from Base is selected, its style will be the same as the view that it was projected from. To override the style from the view that it was created from, uncheck Style from Base and then pick the desired style.

Scale: Set the scale of the view relative to the part or assembly. Type in a scale in the box or pick the down arrow to pick from a list of predefined scales. If Scale from Base is selected, you cannot change the scale of a dependent view. To see the scale label in the view, check the Show Scale box. To override the scale for a dependent view, uncheck the Scale from Base box and then enter a new Scale.

Options: Uncheck the Hatching box to turn off the visibility of the hatch pattern in the view. To remove the alignment constraint for an orthographic or auxiliary view, uncheck the Align to Base box. To realign a view, right click in the unaligned view and pick Alignment from the context menu and choose the alignment type and then pick the view to align to. To turn off the visibility of the view projection lines for a section or auxiliary view, uncheck the Definition in Base View box.

Reference Data Display: In this section the display of reference data will be set.

DELETING DRAWING VIEWS

To delete a drawing view either right click in the bounding area of the drawing view or on its name in the browser and pick Delete from the Context menu. Or click in the bounding area of the drawing view and press the DELETE key on the keyboard. A dialog box will appear asking to confirm the deletion of the view. If the selected view has a view(s) that is dependent on it, you will be asked if the dependent views should also be deleted. By default the dependant view will be deleted. To exclude a dependant view from the delete operation, expand the dialog box and click on the Delete cell for the view.

DIMENSIONS

Once the drawing view(s) have been created, you may need to change the value of a model dimension. If the dimensions were not displayed when the view was created, you may want to get the model dimensions so they appear in a view, hide certain dimensions, add drawing (reference) dimensions, or move dimensions to a new location. All of these operations will be covered in the next section.

DIMENSION VISIBILITY

After creating the drawing view, not all the model dimensions may automatically appear. To get the model dimensions to appear for a specific drawing view, right click in the bounding area of a drawing view and pick Get Model Dimensions from the context menu as shown in Figure 4.36, and the appropriate dimensions for that view that were used to create the part will appear. To hide all the model or drawing dimensions for a drawing view, right click in the bounding area of a drawing view and pick Hide Dimensions and from the sub menu pick either Hide Model Dimensions or Hide Drawing Dimensions as shown in Figure 4.37. To turn on the visibility of all model or drawing dimensions for a drawing view that have been turned off, right click in the bounding area of a drawing view and pick Hide Dimensions and from the sub menu either uncheck Hide Model Dimensions or uncheck Hide Drawing Dimensions.

Figure 4.36

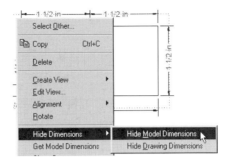

Figure 4.37

DIMENSIONS VALUE AND APPEARANCE

Changing Model Dimensions Value

While working in a drawing view, you may find a need to change the model (parametric) dimensions of a part. You can either open the part file, change the dimensions value, or save the part, and then the change will be reflected in the drawing views. Or change a dimension's value in the drawing view by right clicking on the dimension and pick Edit Model Dimension from the context menu as shown in Figure 4.38. The Edit Dimension text box will appear; type in a new value and then pick the check mark or press ENTER. The associated part will be updated and saved and then the associated drawing view(s) will automatically be updated to reflect the new value.

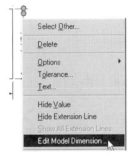

Figure 4.38

DRAWING (REFERENCE) DIMENSIONS

After laying out the drawing views, you may find that another dimension is required to better define the part. You could go back and add a parametric dimension to the sketch if the part was unconstrained, a driven dimension if the sketch was fully constrained, or add a drawing (reference) dimension to the drawing view. A drawing dimension is not a parametric dimension, but it is associative to the geometry that it is referenced to. The drawing dimension reflects the length of the geometry being dimensioned. After a drawing dimension is created and the value of the geometry that was dimensioned changes, the drawing dimension will get updated to reflect the change. A drawing dimension is added with the General Dimension tool from the Drawing Annotation Panel Bar, as shown in Figure 4.39. Create a drawing dimension in the same manner you would a parametric dimension.

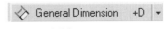

Figure 4.39

AUTO BASELINE DIMENSIONS

Another option to add multiple drawing (reference) dimensions to a drawing view in a single operation is to use the Baseline Dimension tool. The Baseline Dimension tool will place a baseline dimension about the selected geometry. The dimensions can be either horizontal or vertical. Once the dimensions are placed, they are grouped together into a set and members can be added and deleted from the set, a new origin can be placed and the dimensions can automatically be rearranged to reflect the change.

Creating Auto Baseline Dimensions
To create automatic baseline drawing dimensions, follow these steps:

- Pick the Baseline Dimension tool from the Annotation toolbar by picking the down arrow of the General Dimension tool and then pick Baseline Dimension as is shown in Figure 4.40.

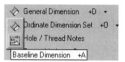

Figure 4.40

- Define the origin line by selecting an edge.
- Draw a box around the geometry that you want to dimension. Move the mouse into position where the first point of the box will be. Press and hold down the left mouse button and move the mouse so the preview box encompasses the geometry that you want to dimension and then release the mouse button.
- When you are done selecting geometry, right mouse click and pick Continue from the context menu.
- Then move the mouse to position where the dimensions will be placed. When the dimensions are in the correct place pick the point with the left mouse button.
- To complete the operation either press the ESC key or right click and pick Done from the context menu.

Editing Auto Baseline Dimensions
After creating dimensions using the Baseline Dimension tool, there are multiple edits that can be performed beyond the basic dimension options. To perform these edits move the mouse over the baseline dimension and small green circles will appear on the dimensions like what is shown in Figure 4.41. Follow the sequence for the following options.

Delete: This will delete *all* the baseline dimensions that are in the set. With the green circles on the screen, right click and pick Delete from the context menu.

Arrange: Will rearrange the dimensions to the spacing that is determined through the Dimension Style.

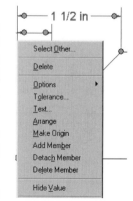

Figure 4.41

Make Origin: This option will change the origin of the baseline dimensions. With the green circles on the screen, select an edge that will define the new baseline and then right click and select Make Origin from the context menu.

Add Member: This option will add a drawing dimension to the baseline dimensions. After creating a drawing dimension move the mouse over the baseline dimensions and with the green circles on the screen, right click and pick Add Member from the context menu and then pick the drawing dimension. The drawing dimension will be added to the set of baseline dimensions and rearranged in the proper order.

Detach Member: This option will remove a dimension from the set of baseline dimensions. With the green circles on the screen, right click on a green circle on the dimension that you want to detach and then right click and pick Detach Member from the context menu. The detached dimension can then be moved and edited like a normal drawing dimension.

Delete Member: This option will erase a dimension from the set of baseline dimensions. With the green circles on the screen, right click on a green circle on the dimension that you want to erase and then right click and pick Delete Member from the context menu and the dimension will be deleted.

ORDINATE DIMENSIONS

Another option for creating drawing dimensions is to create ordinate dimensions. There are two methods for creating ordinate dimensions: ordinate dimension sets and ordinate dimension.

Ordinate Dimension Set: All the ordinate dimensions that are created using the Ordinate Dimension Set tool, in a single operation, will be grouped together but can be edited individually or as a set. When creating an Ordinate Dimension Set, the first dimension created will be used as the origin. The origin dimension needs to be a member of the set and can be changed to a different dimension. If the location of the origin changes or the origin member changes, the other members will be updated to reflect the new location.

Ordinate Dimension: Ordinate dimensions created with the Ordinate Dimension tool are recognized as individual objects and an origin indicator will be created as part of the operation. If the origin indicator location is moved, the other ordinate dimensions will be updated to reflect the change.

Both methods will create drawing dimensions that reference the geometry and will be updated to reflect any changes in the geometry that they are dimensioned to. Ordinate dimensions can be placed on any point, center point, or straight edge. Before adding ordinate dimensions to a drawing view, create or edit a dimension style to reflect your standards. When you place ordinate dimensions, they will automatically be aligned to avoid interfering with other ordinate dimensions.

To create an Ordinate Dimension Set:

1. Create a drawing view.
2. Pick the Ordinate Dimension Set tool from the Annotation Panel Bar as shown in Figure 4.42.

   Figure 4.42

3. Pick the desired dimension style from the Style list in the Command Toolbar.
4. Pick a point to set the origin for the dimensions.
5. Pick a point on the drawing view to locate the dimension and then move the mouse to place the dimension in a horizontal or vertical orientation. Click to place the dimension.
6. Continue selecting points and placing dimensions that will make up the dimension set.
7. To change options for the dimension set, right click at any time and then select Options.
8. To create the dimensions, right click and pick Create from the context menu.
9. To edit the origin, right click on the dimension that will be the origin and pick Make Origin from the context menu. The other dimension values will be updated to reflect the new origin.
10. To edit a dimension set, right click on a dimension in the set and select the desired option from the context menu.

To create an Ordinate Dimension:

1. Create a drawing view.
2. Pick the Ordinate Dimension tool from the Annotation Panel Bar as shown in Figure 4.43.
3. Pick the desired dimension style from the Style list in the Command Toolbar.

   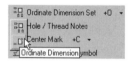

   Figure 4.43

4. In the graphics window, select the view to dimension.
5. Pick a point to set the origin indicator for the dimensions.
6. Pick a point to locate the dimension and then move the mouse to place the dimension in a horizontal or vertical orientation. Click to place the dimension.
7. Continue selecting points and placing dimensions.
8. To create the dimensions, right click and pick Create from the context menu; you can then continue placing ordinate dimensions in another orientation.

9. To create the dimensions and end the operation, right-click and select Done from the context menu.

10. To edit a dimension, right click on a dimension and select the desired option from the context menu.

11. To edit an ordinate dimension leader, grab an anchor point (green circle) and drag it to the desired location.

12. To edit the origin indicator, do one of the following:

    - Grab the origin indicator and drag it to the desired location.
    - Double click the origin indicator and enter the precise location in the Origin Indicator dialog box.

Dimension Appearance

After a model or a drawing dimension exists on the drawing, its appearance can be overridden from the setting in the dimension style that it was based on. The change will only affect the dimension that is selected and supersedes the Dimension Style settings. If a dimension style is applied to a dimension that has overrides, the overrides will be lost. The following properties of a dimension can be changed: style, arrowhead placement, leaded visibility, tolerance, text, and extension line visibility. To change a dimension's style, select it and then from the Style drop list, pick the new style. Multiple dimensions can be selected by holding down the CTRL key and picking them. To change a dimension's appearance, move the mouse over the dimension and when the green circles appear on the dimension, press the right mouse button and a context menu will appear as shown in Figure 4.44. Following is a description of the available tools.

Figure 4.44

Delete: Select this option to delete the dimension from the drawing. If it is a part dimension it is NOT deleted from the part, only the drawing view.

Options:

  Arrowhead Inside: When checked, the arrowheads will be placed inside the extension lines; uncheck it for the arrowheads to be placed outside the extension lines.

  Leader: Select this option to turn on the visibility of a leader line that points to the dimensions value.

Tolerance: Select this option to add a tolerance to the dimension. The tolerance will have no effect on the model. The tolerance is for reference in the drawing.

Text: Select this option to change the text height, font, or add text before or after the value.

Hide Value: Select this option to remove the value from the dimension.

Hide Extension Line: Select this option to hide an extension line. Move the mouse over the extension line that will be hidden and then right click and select Hide Extension Line from the menu. To hide the other extension line, repeat the process.

Show All Extension Lines: Select this option to turn on the visibility of both the extension lines for the chosen dimension. Move the mouse over the dimension with the hidden extension line(s) and then right click and select Show All Extension Lines from the menu.

MOVING DIMENSIONS

To move a dimension by lengthening or shortening the extension lines or moving the text, position the mouse over the dimension until the four circles appear similar to what is shown in Figure 4.45. Then press and hold down the left mouse button and move the dimension to a new location. Depending upon where the dimension has been moved, either only the text will move or the text and extension lines will move and stretch. For example if the text in a vertical dimension is moved up and to the right, the text will move up and the extension lines will stretch to the right. To move or reattach the starting point for an extension line, move the mouse over the dimension's text until the four arrows appear and then pick the green circle at the end of the extension line that will be moved and with the mouse button pressed, move the point to a new location. If the point is moved off the geometry, a dialog box will appear warning you of the problem. An example of when you would want to reattach a dimension would be in a case where the extrusion dimension appears in the middle of a part and the extension line should begin at the end of the part.

Figure 4.45

Note: You can only change the extension line placement for dimensions that were created in the drawing. Dimensions that are on the drawing due to the use of the Get Model Dimension command cannot have their extension lines moved because they are tied to the model geometry; however, their text can be moved and the extension lines either lengthened or shorted from the edges of the drawing view.

HOLE TABLES

If the drawing view that you are dimensioning contains holes, you can locate them by placing individual dimensions or create a hole table that will list the location and size of all the holes in a view or just the selected holes. The hole location will be listed in both the X and Y coordinates in respect to a hole datum that will be placed before creating the hole table. If a part contains a hole that was created by extruding a circle, it will not be placed in the hole table; they must be hole features. After placing a hole table, if you add, delete, or move a hole to a part, the hole table will automatically be updated to reflect the change. Each hole in the table is automatically given an alphanumeric tag as its name and can be edited by either double clicking on the tag in the drawing view, in the hole table or right clicking on the tag in the drawing view or in the hole table and picking Edit Tag from the context menu. Then type in a new tag name, and the change will appear in the drawing view and hole table. There are two tools for creating a hole table: Hole Table–Selection and Hole Table–View. The tools are located in the Drawing Annotation Panel Bar as shown in Figure 4.46.

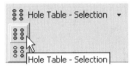

Figure 4.46

Hole Table–Selection will create a hole table from only the selected holes.

Hole Table–View will create a hole table based on all the holes in a selected view.

To create a hole table based on an existing drawing view that shows holes in a plan view, follow these guidelines.

1. Pick the Hole Table tool from the Annotation Panel Bar as shown in Figure 4.46.
2. Pick the view that the hole table will be based on.
3. Pick a point to locate the Origin.
4. If the Hole Table–Selection tool was used, select the holes to include in the hole table, the holes can be selected individually or you can window around the holes.
5. Pick a point to locate the table.
6. The contents of the hole table can be edited by either right clicking on the hole table, the tag name in the hole table or on the tag name in the drawing view. Then choose the desired option from the context menu.

TUTORIAL 4.3—EDITING DRAWING VIEWS AND DIMENSIONING

In this tutorial you will edit existing drawing views, edit existing dimensions, and add reference dimensions.

1. If it is not already open, open the file \Inventor Book\Chapter 4 Drawings.idw.
2. If the first sheet, A Size:1, is not active, make it active by double clicking the sheet's icon.
3. Change the value of the 6.00 inch horizontal dimension to 7 inches by right clicking on the 6 in dimension and select Edit Model Dimension from the

context menu. Type in 7 and pick the check mark in the dialog box or press the ENTER key. The part and drawing views will automatically be updated to reflect the change. *Do not* close the drawing file.

4. Open the part file \Inventor Book\Drawing View Part.ipt and change the 7 in dimension to 5.50 in. Then update the part, save the part file.

5. Close the part file and when the window containing the drawing file is active, you'll see that the 5.50 in dimension is updated in the drawing views.

6. Edit the isometric views and change their display Style to shaded. To edit a view either double click in the view or right click in the view and pick Edit View from the context menu. If Style from Base is checked, it will need to be unchecked before the change can be made.

7. Move the drawing views around until your screen resembles Figure 4.47. To move a view, pick in the view and with the mouse button pressed down, drag the view to a new location.

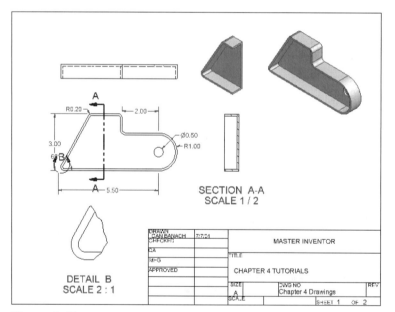

Figure 4.47

8. Practice moving dimensions by dragging them to new locations.

9. When done moving dimensions, hide all the model dimensions in the front view by right clicking in the front view and pick Hide Dimensions>Hide Model Dimensions from the context menu.

10. Practice adding drawing dimensions. Use the General Dimension, Baseline Dimension, Ordinate Dimension Set, and the Ordinate Dimension tools from the Drawing Annotation Panel Bar.

11. When done practicing with drawing dimensions, hide all the drawing dimensions by right clicking in the view(s) with the drawing dimensions and pick Hide Dimensions>Hide Drawing Dimensions from the context menu.

12. Unhide all the model dimensions in the front view by right clicking in the front view and pick Hide Dimensions>Hide Model Dimensions from the context menu.

13. Add the text "TYP" after the R0.20 dimension by right clicking on the dimension and picking Text from the context menu. In the Format Text dialog box, type in the TYP text after the <<>> signs. When finished, the text in the dialog box should look like <<>> TYP. Click OK to apply the change to the text. When done, the front view should resemble Figure 4.48.

14. Practice editing the drawing views, dimensions, and adding reference dimensions.

15. When done, save the file; this file will be used in the next tutorial.

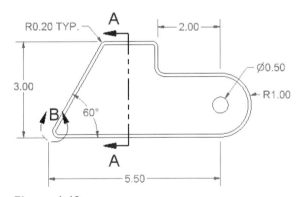

Figure 4.48

ANNOTATIONS

To complete an engineering drawing, you add annotations such as centerlines, surface texture symbols, weld symbols, geometric tolerance symbols, text, bill of materials, and balloons. Before adding annotations to a drawing, make the drawing standard active and make modifications to it as needed.

CENTERLINES

When you need to annotate the centers of holes, circular edges, or the middle (center axis) of two lines, there are four ways to construct the needed centerlines. You use the Center Mark, Center Line Bisector, Center Line, and Centered Pattern tools from the Drawing Annotation Panel Bar as shown in Figure 4.49. The centerlines are associated to the geometry

Figure 4.49

that is selected when it is created. If the geometry changes or moves, the centerlines will automatically update to reflect the change. Following are guidelines for creating the different types of centerlines. To edit a center mark, centerline bisector, centerline or center pattern, position the mouse over it. When the green circles appear right, click and pick from the corresponding Edit options from the context menu.

To add a Center Mark, follow these guidelines.

- Pick the Center Mark tool from the Drawing Annotation Panel Bar.
- In the graphics window, pick the geometry where you want to place a center mark.
- Continue placing center marks by picking geometry.
- To complete the operation, right click and select Done from the context menu.

Note: If the features form a circular pattern, the center mark for the pattern is automatically placed when you have selected all of the members.

To add a Centerline Bisector, follow these guidelines.

- Pick the Centerline Bisector tool from the Drawing Annotation Panel Bar.
- In the graphics window, pick two lines to place the centerline bisector between.
- Right click and select Create from the context menu to place the Centerline Bisector.
- Continue placing Centerline Bisectors by picking geometry.
- To complete the operation, right click and select Done from the context menu.

To add a Centerline follow these guidelines.

- Pick the Centerline tool from the Annotation Panel Bar.
- In the graphics window, pick a piece of geometry for the start of the centerline.
- Click a second of geometry for the ending location.
- Right click and select Create from the context menu to create the centerline. The centerline will be attached to the midpoints of the selected geometry.
- Continue placing Centerlines by picking geometry.
- To complete the operation, right click and select Done from the context menu.

To add a Centered Pattern follow these guidelines.

- Click the Centered Pattern tool from the Drawing Annotation Panel Bar.
- In the graphics window, pick the defining feature for the pattern to place its center mark.
- Click the first feature of the pattern.

- Continue picking features in a clockwise direction until all the features are added to the selection set.
- Right click and select Create from the context menu to create the centered pattern.
- To complete the operation, right click and select Done from the context menu.

SURFACE TEXTURES, WELD SYMBOLS, AND FEATURE CONTROL FRAMES

To add more detail annotations to your drawing, you can add surface texture symbols, weld symbols, and feature control frames by picking the corresponding tool from the Drawing Annotation Panel Bar as shown in Figure 4.50. Then follow these guidelines.

Figure 4.50

- Click the tool from the Drawing Annotation Panel Bar.
- Pick a point where the leader will start at.
- Continue picking points to position the extension lines.
- Right click and select Continue from the context menu.
- Fill in the information as needed in the dialog box.
- When done pick the OK button in the dialog box.
- To complete the operation, right click and select Done from the context menu.
- To edit a symbol, position the mouse over it and when the green circles appear, right click and pick the corresponding Edit option from the context menu.

SKETCHED SYMBOLS

To include a customized symbol on the drawing sheet such as safety symbols or company logos, you will need to define a new symbol. To create a symbol, in the browser under Drawing Resources, right click on Sketched Symbols and pick Define New Symbol from the menu as shown in Figure 4.51. Use the tools from the Sketch toolbar to create the symbol. To save the symbol, right click either in the graphics area or in the browser under Drawing Resources; right click on Sketched Symbols and pick Save Sketched Symbol from the menu. Give the Sketched Symbol a name and click Save in the Sketched Symbol dialog box. The new symbol will not be available to new drawings unless it is saved in a template file. To insert

Figure 4.51

the symbol in the drawing sheet, either double click on the symbols name in the Browser under Drawing Resources>Sketched Symbols or right click on the symbols name and pick Insert from the menu. The symbol can also be inserted with an arrow by picking Insert Callout from the context menu.

TEXT

To add text to the drawing, pick either the Text or Leader text tool from the Drawing Annotation Panel Bar as shown in Figure 4.52. The Text tool will add text while the Leader text tool will add a leader and text. When placing the text, in the Format Text dialog box pick the orientation and text style as needed and then type in the text and pick the OK button to place the text. To edit the text or text leader position, move the mouse over it and when the green circles appear, right click and pick the corresponding Edit option from the context menu.

Figure 4.52

HOLE AND THREAD NOTES

Another annotation that can be added is a hole or thread note. Before a hole or thread note can be placed in a drawing, a hole or thread feature must exist (extruded circles cannot be annotated as a hole) as well as a drawing view that shows the hole in a plan view. If the hole or thread feature changes, the note will automatically be updated to reflect the change.

To create a hole or thread note follow these guidelines.

- Pick the Hole / Thread Notes tool from the Drawing Annotation Panel Bar as shown in Figure 4.53.

Figure 4.53

- Select the hole or thread feature that will be annotated.
- Pick a second point to locate the leader and the note.
- To complete the operation, right click and select Done from the context menu.
- To edit a note, position the mouse over it and when the green circles appear, right click and pick the corresponding Edit option from the context menu.

PARTS LISTS

The final step in creating a drawing is usually to generate a parts list (bill of material). There are numerous company requirements for creating parts lists. This section will highlight the functionality of the options of the parts list tools in Inventor. Practice different techniques and develop your own parts list standards. The parts list that Inventor creates is associative to the part(s) that the parts list is generated from. The file properties will help propagate the appropriate fields in the parts list. In an assembly, the quantity of the parts will automatically be updated in the parts list. Before creating a parts list, establish how the parts list will appear by adjusting the parts list standards. You can choose from the following standards: ANSI, BSI, DIN, GB, ISO, JIS, or you can create your own standard based on one of these standards. From the Format pull down menu pick Standards and then select the Parts List tab in the Drafting Standards dialog box as shown in Figure 4.54. Adjust the settings as needed, changes will only appear in the active file. Save the changes to a template file if new drawing files will utilize the same settings.

To create a parts list follow these guidelines.

- Pick the Parts List tool from the Drawing Annotation Panel Bar as shown in Figure 4.55.
- Pick a point in the view to base the parts list on.
- Place the parts list in the drawing. The information in the parts list is populated from the properties of each component.

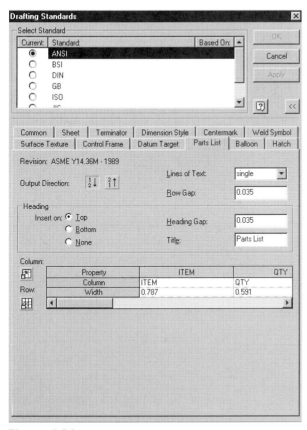

Figure 4.54

Figure 4.55

If after creating the parts list you need to alter its appearance or contents, you can either double click on the parts list or move the mouse over the parts list and a small green dot will appear at one of its vertices. Right click and select Edit Parts List from the context menu. Then the Edit Parts List dialog box will appear as shown in Figure 4.56. Select the tool that you need and follow the options. Following is a brief explanation of the tools that are available.

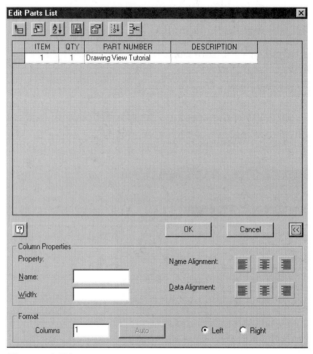

Figure 4.56

PARTS LIST OPERATIONS (TOP ICONS)

Compare: Select the compare tool to compare the parts list to the current properties for the parts. Values that are inconsistent are highlighted in the spreadsheet view. To resolve, right click a highlighted cell and select the desired resolution from the menu.

Column Chooser: Opens the Parts List Column Chooser dialog box where you can add, remove, or change the order of the columns for the selected parts list. Data for these columns is populated from the Properties of the component files.

Sort: Opens the Sort Parts List dialog box where you can change the sort order for items in the selected parts list.

Export: Opens the Export Parts List dialog box so that you can save the selected parts list to an external file with a file type of your choice.

Heading: Opens the Parts List Heading dialog box where you can change the title text or heading location for the selected parts list.

Renumber: Use this tool to change item numbers of parts in the parts list.

Add Custom Parts: Select this tool to add a row to the parts list and then enter the custom parts information. To remove a custom row, highlight the row and press the delete key on your keyboard, or right click within the row and select Remove. This is a useful option for displaying parts in the parts list that are not components or graphical data such as paint or a finishing process.

SPREADSHEET VIEW

This is the middle portion of the Edit Parts List dialog box where the contents of the selected parts list is shown. As you make changes to the parts list properties, the changes are reflected in the spreadsheet. Items that have corresponding balloons are marked with a balloon symbol on the left hand column.

MORE SECTIONS OF THE PARTS LIST DIALOG

Column Properties
Name: Displays the name of the selected column. If needed enter a new name.

Width: Sets the width of the selected column. Enter the width in the box.

Name Alignment: Sets the alignment for the column title. Click a button to select left, center, or right alignment.

Data Alignment: Sets the alignment for the data in the column. Click a button to select left, center, or right alignment.

Note: The data in the parts list is populated from the properties of each component or subassembly. Data input in the parts list is not written back to the associated component or subassembly.

CREATING AN AUTOCAD 2-D DRAWING FROM INVENTOR DRAWING VIEWS

After creating drawing views with Inventor, you may need to output the drawing views so someone with AutoCAD can work with them. To export an AutoCAD 2-D "DWG" or "DXF" file from Inventor drawing views, select Save Copy As from the File pull down menu. Then from the Save Copy As dialog box choose the file format to be exported and pick the Options button in the dialog box to have a wizard guide you through the options for creating the new file. A new file will be generated that will contain all 2-D objects. This 2-D file will have no relationship back to the Inventor part, assembly or drawing file that it was created from.

TUTORIAL 4.4—ANNOTATIONS

In this tutorial you will complete the drawing by adding annotations.

1. If it is not already open, open the file \Inventor Book\Chapter 4 Drawings.idw.
2. If the first sheet A Size is not active, make it active by double clicking on the sheets icon in the browser.
3. Place center marks at the centers of the hole and circular edges in the front and detail view by picking the Center Mark tool from the Drawing Annotation Panel Bar. In the graphics window, pick the hole, circular edges, or the center point to place the center marks then select Done from the context menu. When finished, the front and detail views should resemble Figure 4.57.

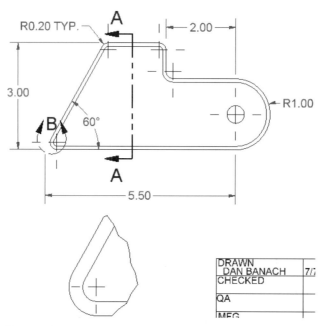

Figure 4.57

4. Create a hole note by picking the Hole / Thread Notes tool from the Drawing Annotation Panel Bar and select the hole on the right side of the front view. Pick a second point to locate the leader down and to the right like what is shown in Figure 4.58. To complete the hole note operation, right click and select Done from the context menu.
5. Place a surface texture symbol on the front edge of the right view by picking the Surface Texture Symbol tool from the Drawing Annotation Panel Bar and pick a point on the top of the front edge. Then pick another point down and to the left and then right click and select Continue from the context menu.

Type in 32 as the Roughness Average–maximum value, and when done pick the OK button in the dialog box. To complete the operation, right click and select Done from the context menu and the hole note and surface texture symbol should resemble Figure 4.58.

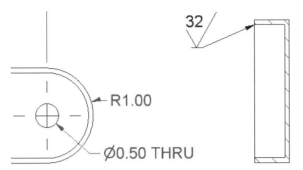

Figure 4.58

6. To get accurate information in the parts list, you will modify the properties of the file that the views are generated from. Open the part file \Inventor Book\Drawing View Part.ipt and from the file pull down menu pick Properties. In the Properties dialog box click the Project tab and change the Part Number to COVER 123 and Description to HOT ROLL STEEL as shown in Figure 4.59 and when done pick the OK button.

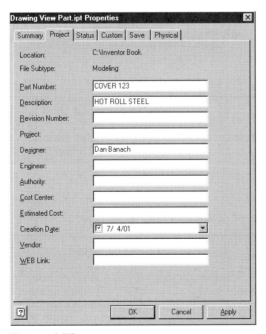

Figure 4.59

7. Save and close the part file.

8. In the drawing file pick the Parts List tool from the Drawing Annotation Panel Bar and then pick a point in the front view to base the parts list on. Place the parts list in the upper right corner of the drawing border. The information in the parts list is populated from the properties of each component.

9. Add a drawing dimension in the right side view to depict its depth and in the detail view to depict the wall thickness. If the drawing views are touching the parts list, move them. When you are done, your drawing should resemble Figure 4.60.

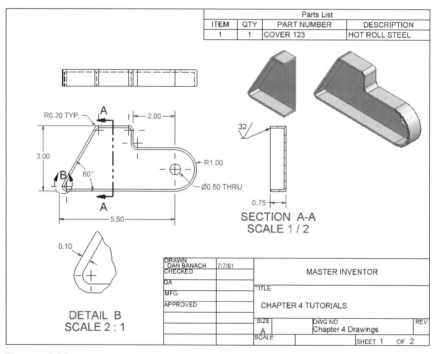

Figure 4.60

10. Save and then close the file.

REVIEW-SEQUENCE FOR CREATING DRAWING VIEWS

Following is a brief outline reviewing the steps that are needed to create a new sheet with drawing views.

1. Create a New Sheet or select an existing drawing sheet from the browser.

2. Add a border.

3. Add a title block.

4. Creating drawing views based on a part, assembly, or presentation.
5. Get Model Dimensions if not done in conjunction with step 4.
6. Clean up drawing views.
7. Add additional dimensions and annotations.
8. Add a parts list.

PRACTICE EXERCISES

The following exercises are intended to challenge you by providing problems that are open-ended. As with the kinds of problems you'll encounter in work situations, there are multiple ways to arrive at the intended solution. In the end, your solution should match the drawing views that are shown. Follow the directions for each of the exercises. When you have completed the part, save the file.

Exercise 4.1—Drain Plate

Start a new drawing file based on the default Standard.idw file. Then create the drawing views as detailed in Figure 4.61. In this exercise, base the views on the file that you created

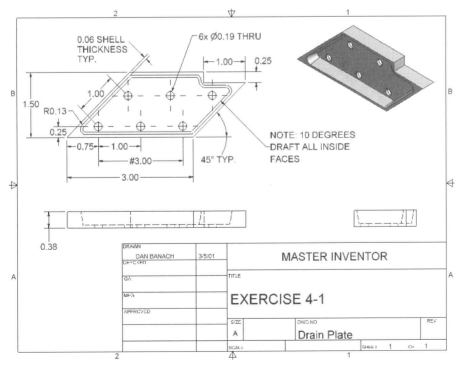

Figure 4.61

in the chapter 3 exercise: \Inventor Book\Drain Plate.ipt. Change the sheet size to "A" and create the views at full scale. Adjust the file properties as needed and when done save the file as \Inventor Book\Drain Plate.idw.

Exercise 4.2—Connector

Start a new drawing file based on the default Standard.idw file. Then create the drawing views as detailed in Figure 4.62.

 Note: For clarity the text height is shown larger. In this exercise base the views on the file that you created in the chapter 3 exercise: \Inventor Book\Connector.ipt. Change the sheet size to "A" and create the views at full scale. Adjust the file properties as needed and when done save the file as \Inventor Book\Connector.idw.

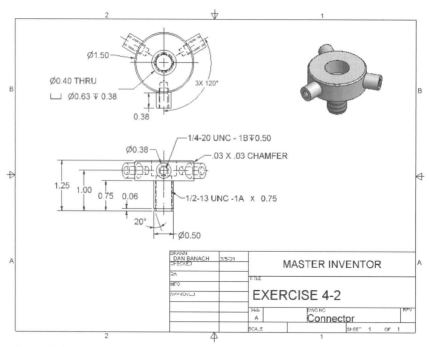

Figure 4.62

REVIEW QUESTIONS

1. A drawing can have an unlimited number of sheets.   True or False?

2. A drawing's sheet size is normally scaled to fit the size of the drawing views.   True or False?

3. There can only be one base view per sheet.   True or False?

4. An isometric view can only be projected from a base view.   True or False?

5. A broken view can only be derived from a base view.   True or False?

6. Drawing dimensions can parametrically drive dimensional changes back to the part.   True or False?

7. Explain how to shade an isometric drawing view.

8. Ordinate dimensions are for reference only and cannot parametrically change the parts size.   True or False?

9. When creating a hole note using the Hole / Thread Notes tool, circles that are extruded to create a hole can be annotated.   True or False?

10. Exported AutoCAD 2-D drawing views of Inventor drawing views are fully associative back to the Inventor drawing file.   True or False?

CHAPTER 5

Assemblies

In the first three chapters you learned how to create a component in its own file. In this chapter you will learn how to place individual component files into an assembly file. This process is referred to as *bottom up* assembly modeling. You also learn to create components in the contents of the assembly file, which is referred to as *top down* assembly modeling. After creating components, you will then learn to constrain the components to one another using assembly constraints, edit the assembly constraints, check for interference, and learn how to create presentation files that show how the components are assembled or disassembled.

CHAPTER OBJECTIVES

After completing this chapter, you will be able to:

- understand the assembly options.
- create bottom up assemblies.
- create top down assemblies.
- create adaptive features.
- create subassemblies.
- constrain components together using assembly constraints.
- edit assembly constraints.
- check for interference.
- replace a component with another component.
- create a presentation file.
- create balloons.
- create a bill of material of an assembly.

CREATING ASSEMBLIES

As you previously learned, component files have the extension of IPT and a component file can only have one component in it. In this chapter you will learn how to create assembly files (file extension IAM). An assembly file holds the information that is needed to assemble the components together. All the components in an assembly are referenced in, meaning that each component exists in its own component IPT file and its definition is linked into the assembly. The components in an assembly can be edited while in the assembly or you can open the component file and edit it. When the changes are made to a component and the component is saved, the changes will be reflected in the assembly after the assembly is opened or updated. There are three methods for creating assemblies: bottom up, top down, and assemblies that are a combination of both top down and bottom up techniques. Bottom up refers to an assembly where all the components were created in individual component files and are referenced into the assembly. A top down approach refers to an assembly where the components are created from within the context of the assembly. In other words, the user creates each component from within the top level assembly. Each component in the assembly is saved to its own component (IPT) file. Both the bottom up and top down assembly techniques will be covered in the following sections.

To create a new assembly file select the New icon from the What To Do section and then select the Standard.iam icon from the Default template tab as shown in Figure 5.1. There are two other methods for creating a new assembly: you can

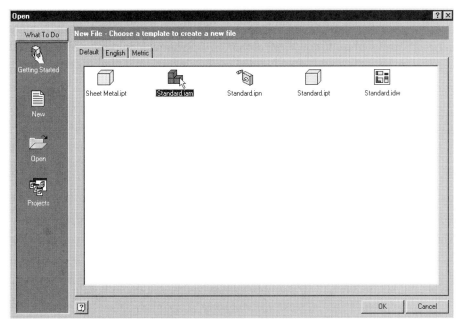

Figure 5.1

select New from the File pull down menu and then select the Standard.iam icon or from the left side of the Standard toolbar select the down arrow of the New icon and select Assembly. After issuing the new component operation, Inventor's tools will change to reflect the new assembly environment. In the panel bar the assembly tools will be shown. The assembly tools will be covered throughout this chapter.

 Note: There is no correct or incorrect way to create an assembly. Based on experience you will determine what method works best for the assembly that you are creating.

An assembly can be created using both the bottom up and top down assembly techniques.

No matter if you placed or created the components in the assembly, all the components will be saved out to their own individual IPT files and the assembly will be saved as an IAM file.

ASSEMBLY OPTIONS

Before creating an assembly let's examine the assembly option settings. Under the Tools pull down menu, select Application Options and the Options dialog box will appear; select the Assembly tab and your screen should look similar to Figure 5.2.

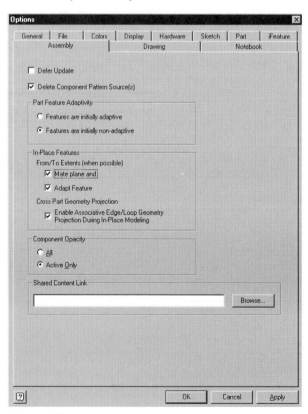

Figure 5.2

Following is a description of the Assembly options. These settings are global and will affect how new components are created or placed in the assembly.

Defer Update: When checked, the assembly will automatically update to reflect the change to a component. If unchecked and a component is changed that will affect the assembly, you will have to manually pick the Update button to get the assembly up to date.

Delete Component Pattern Source(s): When checked, the source component will automatically be deleted when pattern elements are deleted. If unchecked, the source component will not be deleted when the pattern elements are deleted.

PART FEATURE ADAPTIVITY

Under this section you will indicate if new features will be adaptive or non-adaptive when they are created. The adaptivity of a feature can also be changed after it has been created. Adaptivity means that a feature will be able to change its size according to the relationship it has with another component. Adaptivity will be covered later in this chapter.

Features Are Initially Adaptive: Select this option if you want under-constrained sketches and features to get their size by determining a relationship to another component in the assembly.

Features Are Initially Non-adaptive: Select this option if you do not want under-constrained sketches and features to get their size from another component in the assembly.

IN-PLACE FEATURES

From/To Extents (when possible)
When creating a component in the context of an assembly, you can determine how the new features will terminate. Multiple options can be selected at the same time.

Mate Plane: Select this option when you to create a new component and have a mate constraint applied to the plane on which it was constructed, but it will *not* be adaptive.

Adapt Feature: Select this option when you want to create a new component and have it adapt to the plane on which it was constructed.

Cross Part Geometry Projection
Enable Associative Edge/Loop Geometry Projection During In-Place Modeling: When checked and geometry is projected from another part onto the active sketch, the projected geometry is associative (sketch associativity) and will update when changes are made to the parent part. Projected geometry can be used to create a sketched feature.

Component Opacity

In this section you will determine if all or only the active component will be opaque when an assembly cross-section is displayed.

All: When checked all the components in the assembly will be opaque.

Active Only: When checked only the active component in the assembly will be opaque; the others will be dimmed.

SHARED CONTENT LINK

Sets the URL for third-party content added to Autodesk Inventor assemblies using the Place Content option on the Insert menu or from the Assembly Panel Bar. Allowable file types are .htm, .html, and .exe.

BOTTOM UP APPROACH

The bottom up assembly approach uses files that are referenced to an assembly file. To create a bottom up assembly, create the components in their individual files. If an assembly file is placed into another assembly, it will be brought in as a subassembly. Any drawing views that were created for these files will not be brought into the assembly file. You will want to make sure that the path(s) for the file location(s) of the placed component(s) is represented in the project file that the assembly is based on. Otherwise the component will not be located when the assembly is reopened. To insert a component into the current assembly, pick the Place Component tool from the Assembly Panel Bar as shown in Figure 5.3 or press the hot key P or right mouse click and select Place Component from the context menu

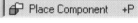

Figure 5.3

and the Open dialog box will appear as shown in Figure 5.4. Then you pick a point in the assembly where you want the component located. If multiple occurrences of the component are needed in the assembly, continue picking points. When done, press the ESC key or right click and select Done from the context menu.

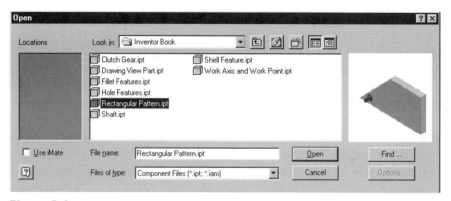

Figure 5.4

TOP DOWN APPROACH

While in an assembly you can create new components. This method is referred to as the top down approach. To create a new component in the current assembly, pick the Create Component tool from the Assembly Panel Bar as shown in Figure 5.5 or right mouse click and select Create Component from the context menu and the Create In-Place Component dialog box will appear as shown in Figure 5.6. Type in a new name, file type, location it will be saved, template file to base it on, and select if the component will be constrained to a face on another component. If the "Constrain sketch plane to selected face or plane" is checked, a flush constraint will be applied between the selected face and the new plane. If it is unchecked, no constraint will be applied, but you will still select a face to start sketching the new component on.

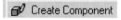

Figure 5.5   Figure 5.6

OCCURRENCES

An occurrence is a copy of an existing component and has the same name as the original component, but the number after the colon will be sequenced. For example, if the original component has a name Bracket.ipt:1, the occurrence will be Bracket.ipt:2. If the original component or an occurrence of the component changes, all the components will reflect the change. To create an occurrence, you can place the component multiple times, or if the component already exists, you can select the component's icon in the Browser and drag an occurrence into the assembly. The copy-paste methods can also be used to place components by right clicking on the component name in the Browser or on the component itself and then right click and pick Copy from the context menu. Then either right click and select Paste from the context menu or press the CTRL and V keys together. If you want an occurrence of the original component to have no relationship to the original component, you will need to make that component active and then issue the Save Copy As tool from the File pull down menu and type in a new name. This new component will have no relationship to the original component and can be placed into the assembly.

MULTIPLE DOCUMENTS AND DRAG AND DROP COMPONENTS

In Inventor you can open as many files as needed. This is referred to as a multiple document environment. The screen can be split as needed to show all the files that are open by using the options under the Window pull down menu. You can switch between the open files to model, edit, or interrogate the files as needed. With the multiple documents open, you can drag a component from one file into an assembly. To do so, open both the assembly and the component file and split the screen so both files are visible. Then with the left mouse button, select the component's name in the Browser and drag it into the assembly file. Then release the mouse button.

ACTIVE COMPONENT

To edit a component while in the assembly, you need to be using the active component. Only one component in the assembly can be active at a time. To make a component active, you can double click on the component in the graphics area; you can double click on the file name or icon in the Browser; or you can select the component name in the Browser with the right mouse button and select Edit from the context menu. Once the component is active, the other component names will be grayed out as shown in Figure 5.7, and the other components in the assembly will be dimmed if Component Opacity, from the Application Options under the Assembly tab, is set to Active Only. Then edit the component as you learned in the previous chapters and save the changes by using the save tool. Since the component is active, only that component will be saved. To make the assembly active, either select the Return button from the Command Bar as shown in Figure 5.8, double click on the assembly name in the Browser, or right click in the graphics window and select Finish Edit from the context menu.

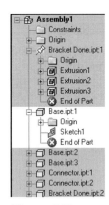

Figure 5.7

Figure 5.8

OPEN AND EDIT

Another method to edit a component in the assembly is to open the component in another window. This can be done by selecting the Open tool from the Standard toolbar, select Open from the File pull down menu, or right click on the component's name in the Browser and Select Open from the context menu. The component will appear in a new window; edit the component as needed and then save the changes. Then activate the assembly file and the changes will appear in the assembly.

GROUNDED COMPONENTS

When assembling components together, you may want to have a component or multiple components that are grounded (stationary), meaning that they will not move. When applying assembly constraints, the other components will be moved to the grounded component(s). By default the first component in the assembly is grounded. There is no limit to how many components can be grounded. It is recommended that at least one component in the assembly be grounded; otherwise the assembly will be able to move. A grounded component is represented with a pushpin before its name in the Browser. To ground or unground a component, right click on the parts name in the Browser and select or deselect Grounded from the context menu as shown in Figure 5.9.

Figure 5.9

SUBASSEMBLIES

While working, you may want to group components together into a subassembly. There are two methods to create a subassembly. Either place an existing assembly into another assembly or create a new assembly from within the assembly. To create a subassembly from within the assembly, issue the Create Component tool then change the file type to "Assembly" as shown in Figure 5.10. Then any component that is placed or created will be a component of the subassembly. To make a subassembly active, double click on the subassembly's name in the Browser. To make the top level assembly active, pick the Return tool from the Command Bar or double click on the top level assembly's name. When working in the top level of the assembly, any subassembly, when selected, will act as a single component. Components can also be promoted into a subassembly or demoted from a subassembly by selecting its name in the Browser and select either Promote or Demote from the context menu.

Figure 5.10

REORDER AND RESTRUCTURE

After components are in an assembly, the order that they appear can be reordered and components can also be restructured from the main assembly to another subassembly or from subassembly to subassembly or subassembly to the main assembly. To either reorder or restructure a component, pick the component's name in the Browser that will be reordered or restructured and with the left mouse pressed move the component up or down in the browser. A line will appear that shows where it will be placed. Keep moving the mouse until the line is positioned in the correct place then release the left mouse button. Figure 5.11 shows how the Browser looks when a component is being restructured into a subassembly. When restructuring components into or from a subassembly, a dialog box may appear stating that assembly constraints may be lost. After completing the restructure, examine the assembly constraints to verify that they are still valid.

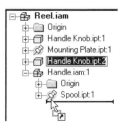

Figure 5.11

TUTORIAL 5.1—BOTTOM UP ASSEMBLY TECHNIQUE

In this tutorial you will create an assembly by using the bottom up assembly technique. You will place components that have already been created. Later in the chapter you will apply assembly constraints.

1. Start a new Assembly file based on the default Standard.iam file.
2. Save the assembly as \Inventor Book\Reel.iam.
3. Pick the Place Component tool from the Assembly Panel Bar. In the Open dialog box select the file \Inventor Book\Reel Mounting Plate.ipt. Then select the *open* button in the Open dialog box.
4. In the graphics area, one occurrence of the component will be placed. Complete the operation by either pressing the ESC key or right click in the graphics area and select Done from the context menu.

5. Pick the Place Component tool from the Assembly Panel Bar. Then in the Open dialog box select the file \Inventor Book\Reel Spool.ipt. Then select the *open* button in the Open dialog box.

6. In the graphics area, pick a location for the spool. Only place one occurrence, complete the operation by either pressing the ESC key or right click in the graphics area, and select Done from the context menu.

7. Pick the Place Component tool from the Assembly Panel Bar. Then in the Open dialog box select the file \Inventor Book\Reel handle Knob.ipt. Then select the *open* button in the Open dialog box.

8. In the graphics area, pick a location for the spool. Only place one occurrence, complete the operation by either pressing the ESC key or right click in the graphics area, and select Done from the context menu. When you are done placing the three components your screen should resemble Figure 5.12.

9. Try moving the components by selecting them with the left mouse button and keep the button depressed and move the mouse. All the components will move except the Reel Mounting Plate because it is grounded. Move the components until your screen resembles Figure 5.12.

10. Save the file; later in the chapter you will add assembly constraints.

11. Close the assembly file.

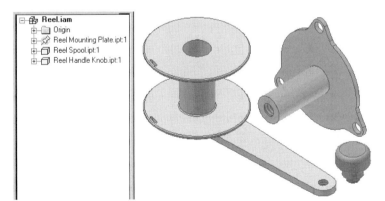

Figure 5.12

TUTORIAL 5.2—TOP DOWN ASSEMBLY TECHNIQUE

In this tutorial you will create an assembly by using the top down assembly technique. You will create two components while in the assembly file and edit them using two different techniques. Later in the chapter you will apply assembly constraints.

1. Start a new Assembly file based on the default Standard.iam file.
2. Pick the Create Component tool from the Assembly Panel Bar. From the Create In-Place Component dialog box type in "Long Link" for the New File Name and then select OK.
3. Sketch, constrain, and dimension the geometry as shown in Figure 5.13. Apply a fix constraint and use equal constraints to fully constrain the sketch.

Figure 5.13

4. Change to an isometric view by selecting the right mouse button while in the graphics area and then select Isometric View from the context menu.
5. Extrude the sketch .1 inch in the default direction. For the Profile select in the middle of the sketch so the circles will not be included in the extrude operation.
6. Save the component file. Since the component is active, only the component file will be saved and not the assembly file.
7. Make the top level assembly active by either picking the Return tool from the Command Bar or double click on the top level assembly's name in the browser.
8. Save the assembly as \Inventor Book\Linkage.iam
9. Pick the Create Component tool from the Assembly Panel Bar. From the Create In-Place Component dialog box type in "Short Link" for the New File Name and check the Constrain sketch plane to selected face box. Pick OK in the dialog box and then select the front vertical face of the Long Link component as shown in Figure 5.14. A flush constraint will be applied and a sketch will become active on the selected plane.

Figure 5.14

10. Sketch, constrain, and dimension the geometry as shown in Figure 5.15. Apply a fix constraint and use equal constraints to fully constrain the sketch.

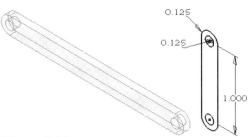

Figure 5.15

11. Extrude the sketch .1 inch in the default direction. For the Profile select in the middle of the sketch so the circles will not be included in the extrude operation.
12. Save the component file. Since the component is active, only the component file will be saved and not the assembly file.
13. Make the top level assembly active by either picking the Return tool from the Command Bar or double click on the top-level assemblies name.
14. Make the component Long Link the active component by either double clicking on the components name in the Browser or double clicking on the component itself.
15. Edit the sketch and change the length to 3 inches as shown in Figure 5.16.

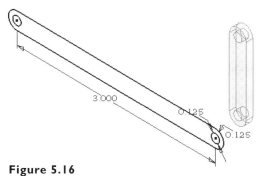

Figure 5.16

16. Update the component and make the top level assembly active by either picking the Return tool from the Command Bar or double click on the top-level assemblies name.

17. You will now edit the component Short Link by opening it and making changes. In the Browser Right click on the component name, Short Link, and select Open from the context menu as shown in Figure 5.17.

18. Edit the sketch and change the length to 1.25 inches as shown in Figure 5.18.

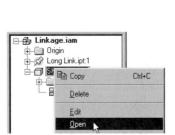

Figure 5.17

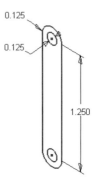

Figure 5.18

19. Update and save the file.
20. Close the Short Link file.
21. The assembly should now be the active file. The change to the Short Link component will appear in the assembly.
22. Now you will place an occurrence of each component. Either select the components' name in the Browser and drag them into the Graphics area or right click on the component name in the Browser or on the component itself and right click and pick Copy from the context menu then either right click and select Paste from the context menu or press the CTRL and V key. When done your screen should resemble Figure 5.19.

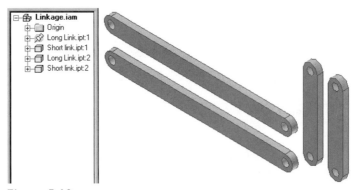

Figure 5.19

23. Save the assembly file and all the dependent component files. Later in the chapter you will add assembly constraints.
24. Close the assembly file.

ASSEMBLY CONSTRAINTS

In the previous sections you learned how to create assembly files, but the components did not have any relationship to each other except for the relationship that was set when a component was created in reference to a face on another component. For example, if a bolt was placed in a hole, and the hole moved, the bolt would not move to the new hole position. Assembly constraints are used to create relationships between components. So, if the hole moves, the bolt will move to the new hole location. In chapter 1, you learned about geometric constraints. When geometric constraints are applied, they reduce the number of dimensions or constraints required to fully constrain a profile. When assembly constraints are applied, they reduce the degrees of freedom (DOF) that allow the components to move freely in space. There are six degrees of freedom: three translational and three rotational. Translational means that a component can move along an axis: X, Y or Z. Rotational means that a component can rotate about an axis: X, Y or Z. As assembly constraints are applied, the number of degrees of freedom will be decreased. Inventor does not need parts to be fully constrained. By default the first component created or added to the assembly will be grounded and will have zero degrees of freedom. More than one component can be grounded as discussed earlier in this chapter. Other components will move in relation to the grounded component(s). To see a graphical display of the degrees of freedom remaining on all the components in an assembly, pick Degrees of Freedom from the View pull down menu as shown in Figure 5.20. An icon will appear in the center of the component that shows the degrees of freedom remaining on the component. The line and arrows represent translational freedom, and the arc and arrows represent rotational freedom. To turn off the degrees of freedom icons, pick Degrees of Freedom from the View pull down menu.

Figure 5.20

Note: Inventor does not require parts to be fully constrained.

Tip: You can turn on the DOF symbols for a single (or a few) component(s) by right clicking on the components name in the browser, selecting Properties from the context menu, then selecting the Degrees of Freedom checkbox on the Occurrence tab. Doing the same steps will toggle the symbols off once they have been turned on.

When placing or creating components in an assembly, it is recommended to have them in the order in which they will be assembled. The order will be important when placing assembly constraints and creating presentation views. If the components are not in the correct order, they can be reordered as was discussed earlier in the chapter. Like sketches, components in assemblies do not need to be fully constrained. When constraining components to one another, you will need to understand the terminology used. A list of terminology that is used with assembly constraints follows.

Line: Can be the centerline of an arc or circular edge, a selected edge or a work axis.

Normal: A vector that is perpendicular to the outside of a planar face.

Plane: Can be defined by the selection of a plane or face: two non-collinear lines, three non-linear points, one line and a point that does fall on the line. When edges and points are used to select a plane, this is referred to as a construction plane.

Point: Can be an endpoint or a midpoint of a line, the center of an arc, or circular edge.

Offset: The distance between two selected lines, planes, points, or any combination of the three.

TYPES OF CONSTRAINTS

Inventor has four types of assembly constraints: Mate, Angle, Tangent and Insert; two types of Motion constraints: Rotation and Rotation-Translation; and a Transitional constraint. The constraints can be accessed through the Place Constraint tool from the Assembly Panel Bar as shown in Figure 5.21 or right mouse click and select Place Constraint from the context menu or use the hot key C. The Place Constraint dialog box will appear as shown in Figure 5.22. The dialog box is divided into four areas. Each will be described. Depending upon the constraint type, the option titles may change.

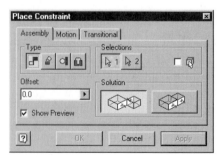

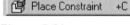

Figure 5.21

Figure 5.22

ASSEMBLY TAB

Type
Select the type of assembly constraint to apply: Mate, Angle, Tangent, or Insert.

Selections
In this area pick the button with the number 1 and then select a component's edge, face, point, etc. to base the constraint type on. Then pick the button with the number 2 and select a component's edge, face, point, etc. to base the constraint type on. By default, the second arrow will become active after you have selected the first input. You can edit an edge, face, point, etc. of an assembly constraint that has already been applied by picking the number button that corresponds to the constraint and then picking a new edge, face, point, etc. While working on complex assemblies you can check the box on the right side of the Selections area, called Pick part first, and then before selecting a component's edge, face, point, etc., you will need to first select the component.

Offset/Angle
Type in or select, from the drop list, a value for the offset or angle.

Solution
In this area you can select how the constraint will be applied; either the normals will be pointing in the same or opposite directions.

MOTION TAB

Type
Select the type of assembly constraint to apply: Rotation or Rotation-Translation

Selections
In this area pick the button with the number 1 and then select a component's edge, face, point, etc. to base the constraint type on and then pick the button with the number 2 and select a component's edge, face, point, etc. to base the constraint type on. By default, the second arrow will become active after you have selected the first input. You can edit an edge, face, point, etc. of an assembly constraint that has already been applied by picking the number button that corresponds to the constraint and then picking a new edge, face, point, etc. While working on complex assemblies, you can check the box on the right side of the Selections area, called Pick part first, and then before selecting a component's edge, face, point etc. you will need to first select the component.

Ratio
Type in or select, from the drop list, a value for the ratio.

Solution
In this area you can select how the constraint type will be applied. The components will either rotate in the same or opposite directions.

TRANSITIONAL TAB
A transitional constraint will maintain contact between the two selected faces. A transitional constraint can be used between a cylindrical face and a set of tangent faces on another part.

Type
Select transitional as the type of assembly constraint to apply.

Selections
In this area pick the First Selection button and select the first face on the part that will be moving. Pick the Second Selection button and then pick a face that the first part will be moving around. If there are tangent faces they will automatically become part of the selected face.

CONSTRAINT TYPES
In this section each of the assembly constraint types will be explained.

Mate: There are three types of mate constraints: plane, line, and point.

> **Mate Plane:** Assembles two components so that the normals on the selected planes will be opposite one another (mate) or pointing in the same direction (flush) when the components are assembled. Select either mate or the flush condition from the Solution area.
>
> **Mate Line:** Assembles to edges or centerlines of arcs and circular edges together.
>
> **Mate Point:** Assembles two points (center of arcs and circular edges, endpoints and midpoints) together.

Angle: You will specify the degrees between the selected planes.

Tangent: Select the tangent type if you need to define a tangent relationship between planes, cylinders, spheres, cones, and ruled splines. At least one of the selected faces needs to be a curve, and the tangency may be inside or outside the curve.

Insert: Select the circular edges of two different components and the centerlines of the components will be aligned and a mate constraint will also be applied to the planes defined by the circular edges. The Insert constraint takes away five degrees of freedom with one constraint. It only works with components that have circular edges. Circular edges define a centerline/axis and a plane.

Rotation: Use the Rotation constraint to define a component that will rotate in relation to another component by specifying a ratio for the rotation between the two components.

Rotation-Translation: Use the Rotation-Translation constraint to define the rotation relative to translation of a second component.

Figures 5.23 through 5.29 will help you better understand how each type of assembly constraint works. Each figure shows components before and after assembly constraints are applied.

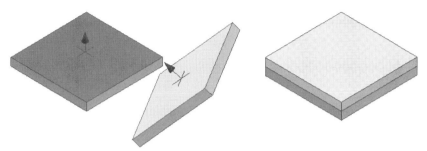

Figure 5.23 *Mate constraint type with mate solution applied.*

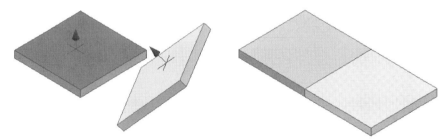

Figure 5.24 *Mate constraint type with flush solution applied.*

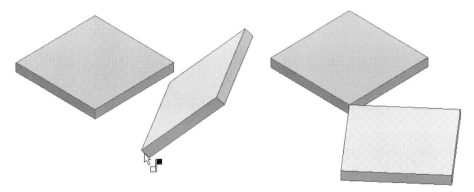

Figure 5.25 *Mate constraint type with two edges selected.*

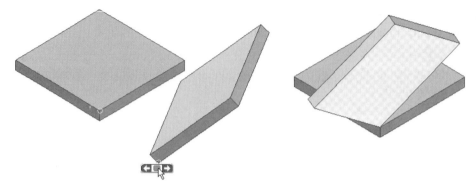

Figure 5.26 *Mate constraint type with two points selected.*

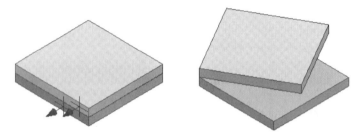

Figure 5.27 *Angle constraint type with two planes selected and 60° angle applied.*

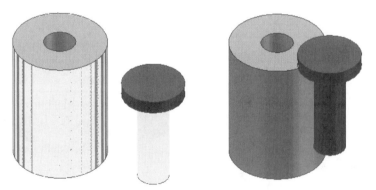

Figure 5.28 *Tangent constraint type with two outside curved faces selected and outside solution applied.*

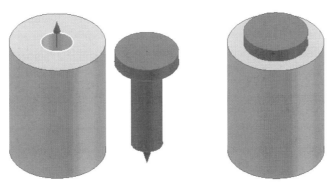

Figure 5.29 *Insert constraint type with two circular edges selected and opposed solution applied.*

APPLYING ASSEMBLY CONSTRAINTS

After selecting the assembly constraint type that you want to apply, the Selections 1 button will become active—or you can select the button if it doesn't. Position the mouse over the face, edge, point, etc. to apply the first assembly constraint. You may need to cycle through the selection set until the correct location is highlighted, then press the left mouse button or the middle rectangle in the cycle dialog. Then position the mouse over the face, edge, point, etc. to apply the second assembly constraint. Again cycle through the selection set until the correct location is highlighted. If the Show Preview option is selected in the dialog box, the components will move to show how the assembly constraint will affect the components, and a snapping sound will be heard when the constraint is previewed. To change either selection, click on the number 1 or 2 button and then select the new input. Type in a value as needed for the offset or angle and pick the correct Solution option until the desired outcome is shown. Select the Apply button to complete the operation.

Snap 'n Go Constraints

Another method to apply an assembly constraint is to hold down the ALT key while dragging a part to another part; no dialog box will appear. The key to dragging and applying a constraint is to select the correct area on the part. Selecting an edge will create a different type of constraint that selecting a face will create. For example, if a circular edge is selected, an Insert constraint will be applied. To apply a constraint while dragging a part, you cannot have another tool active.

To apply an assembly constraint, follow these guidelines.

- Hold down the ALT key, then select the face, edge, etc, on the part that will be constrained.

  Select a planar face, linear edge, or axis to place a mate or flush constraint.

  Select a cylindrical face to place a tangent constraint.

  Select a circular edge to place an insert constraint.

- Drag the part into position. As the part is dragged over features on other parts, the constraint type will be previewed. If the face you need to constrain to is behind another face, pause till the Select Other tool appears. Cycle through the possible selection options, then click the center dot to accept the selection.

- To change the constraint type that is being previewed while the part is being dragged, release the ALT key and enter one of the shortcut keys as follows:

M or **1**: Changes to a mate constraint. Press the spacebar to flip to a flush solution.

A or **2**: Changes to an angle constraint. Press the spacebar to flip angle direction on the selected component.

T or **3**: Changes to a tangent constraint. Press the spacebar to flip between an inside and outside tangent solution.

I or **4**: Changes to an insert constraint. Press the spacebar to flip the insert direction.

R or **5**: Changes to a rotation motion constraint. Press the spacebar to flip the rotation direction.

S or **6**: Changes to a rotation-translation constraint. Press the spacebar to flip the translation direction.

X or **8**: Changes to a transitional constraint.

Note: You can create theoretical or construction planes or edges by moving the mouse cursor over an edge and cycling through the geometry and then moving it to the next edge and cycling through in the same manner until the plane or edge is defined.

A work plane can be used as a plane with assembly constraints.

A work axis can be used to define a line.

EDITING ASSEMBLY CONSTRAINTS

After an assembly constraint has been placed, you may want to edit, suppress, or delete it to reposition the components. There are two methods for editing assembly constraints.

Both methods are through the Browser. In the Browser activate the assembly or subassembly which has the component you want to edit. Expand the component name and you will see the assembly constraints. Double click on the constraint name and an Edit Constraint dialog box will appear, allowing you to edit the constraints. The second method is to right click on the assembly constraint's name in the Browser, and from the context menu choose Edit, Suppress, or Delete as shown in Figure 5.30. If you choose Edit, the Edit Constraint dialog box will appear. If you pick Suppress, the assembly constraint will not be applied and another constraint can be applied or a constraint could be driven if it was being held into position by the assembly constraint. Pick Delete and the assembly constraint will be deleted from the component. If you try to place or edit an assembly constraint and it cannot be applied, an alert box will appear that will explain the problem. Either select new options for the operation or suppress or delete another assembly constraint that is conflicting with it. If an assembly constraint is conflicting with another, a small yellow icon with an exclamation point will appear in the Browser as shown in Figure 5.31. To edit a conflicting constraint in the Browser either double click on its name or right click on its name and pick Recover from the context menu as shown in Figure 5.31. The Design Doctor will appear and will walk you through the steps to fix the problem.

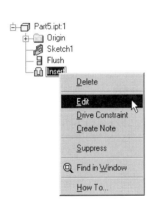

Figure 5.30

Figure 5.31

DRIVING CONSTRAINTS

Mechanical motion can be simulated (driven) using the Drive Constraint tool. To simulate motion either an angle or mate with an offset assembly constraint must exist. Only one assembly constraint can be driven at a time, but equations can be used to create relationships to drive multiple assembly constraints. To drive a constraint, right click on the constraint in the Browser and select Drive Constraint from the context menu as shown in Figure 5.32. The Drive Constraint dialog box will appear similar to Figure 5.33. Depending upon the constraint that you are driving, the units may be different. Enter a Start value; the default value is the angle or offset for the constraint. Enter a value for the End and a pause delay if you want time between the steps. On the top half of the dialog box, you can also choose to create an animation (AVI) file that will show the motion. The AVI file can be replayed without Inventor being installed. To set more details on how the motion will behave, select the << (more) button. Each option of the dialog box will be explained below.

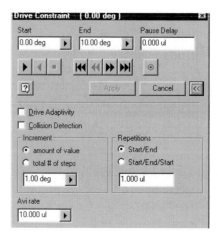

Figure 5.32   **Figure 5.33**

Start: Set the start position of the offset or angle.

End: Set the end position of the offset or angle.

Pause Delay: Set the delay between steps.

Motion and AVI Control Buttons: Use the following buttons to control the motion and to create an AVI file.

- Forward drives the constraint forward.
- Reverse drives the constraint in reverse.
- Stop temporarily stops the constraint drive sequence.
- Minimum returns the constraint to the starting value and resets the constraint driver. This button is not available unless the constraint driver has been run.
- Reverse Step reverses the constraint driver one step in the sequence. This button is not available unless the drive sequence has been stopped.
- Forward Step advances the constraint driver one step in the sequence. This button is not available unless the drive sequence has been stopped.
- Maximum advances the constraint sequence to the end value.
- Record begins capturing frames at the specified rate for inclusion in an animation.

Drive Adaptivity: Check the box if you want the component to adapt while the constraint is driven.

Collision Detection: When checked the constraint will be driven until collision is detected. When interference is detected, the drive constraint will stop and the components where the collision occurs will be highlighted. The constraint value for the collision will also be shown.

Increment: Enter a Value that the constraint will be incremented during the animation.

> **Amount of Value:** Use this option to drive the constraint in this number of increments.
>
> **Total # of Steps:** Use this option to drive the constraint equally per the number of steps.

Repetitions: Choose how the driven constraint will act when it completes a cycle and how many cycles there will be.

> **Start/End:** Drives the constraint from the start value to the end value and resets at the start value.
>
> **Start/End/Start:** Drives the constraint from the start value to the end value and then in reverse to the start value.

AVI Rate: Specify how many frames are skipped before a screen capture is taken of the motion that will become a frame in the completed AVI file.

> **Note:** If you try to drive a constraint and it fails, you may need to suppress or delete another assembly constraint. To reduce the file size of an AVI, reduce the screen size before creating the AVI file and use a solid background color in the graphics window.

INTERFERENCE CHECKING

To check the interference between two sets of stationary components, make the assembly or subassembly where you want to check for interference active. Then pick the Analyze Interference tool from the Tools menu as shown in Figure 5.34. The Interference Analysis dialog box will appear as shown in Figure 5.35. Pick the components that will define the first set. Then pick the Define Set #2 button and choose the components that will define the second set. A component can only exist in one set and a set can contain only a single component. To add or delete components from either set, select the button that defines the set that you want to edit and select components to add to the set or press the CTRL key while selecting a component to remove it from the set. Once the sets are defined, select the OK button. The order in which the components are selected has no significance. If interference is found, an Interference Detected dialog box will appear like what is shown in Figure 5.36. The information in the dialog box defines the x, y, z coordinates of the centroid of the interfering volume and lists the volume of the interference. Also a temporary solid will be created in the graphics window that represents the interference. The interference report can be copied to the clipboard or printed from the tools in the dialog box. When the operation is complete, pick the OK button and the interfering solid will be removed from the screen. After analyzing and finding interference, edit the components to remove the interference.

 Note: Interference can also be detected when driving constraints.

Figure 5.34

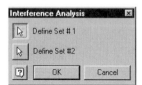

Figure 5.35

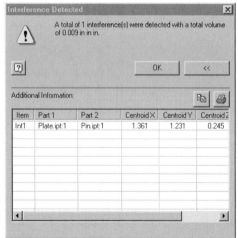

Figure 5.36

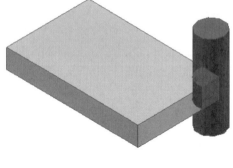

TUTORIAL 5.3—BOTTOM UP ASSEMBLY CONTINUED

In this tutorial you will add assembly constraints, check for interference, create a subassembly, and restructure a component. While working on the assembly, you may need to move the components or dynamically rotate the components to better select the edges. After completing this tutorial, go back and practice other techniques for assembling the components.

1. Open the assembly file \Inventor Book\Reel.iam.
2. Use the Place Constraint tool to place a Mate constraint with "0" offset between faces of the Reel Mounting Plate and the Reel Spool as shown in Figure 5.37. After selecting the faces, pick the Apply button to create the constraint.

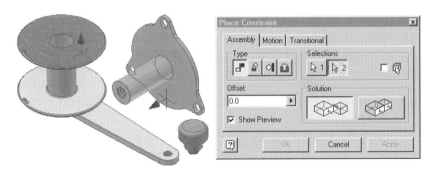

Figure 5.37

3. While still in the Place Constraint dialog box, place a Mate constraint with "0" offset between the centerlines of the Reel Mounting Plate and Reel Spool as shown in Figure 5.38. After selecting the centerlines, pick the Apply button to create the constraint.

 Note: An insert constraint could have been applied instead of independent mate and centerline constraints.

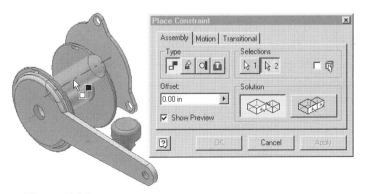

Figure 5.38

4. To see how many degrees of freedom are left on the spool, pick Degrees of Freedom from the View pull down menu. You will see that it can still rotate about the center axis. Pick the Reel Spool and with the left mouse button pressed down, rotate the handle. Remove the degree of freedom symbols from the screen by picking Degrees of Freedom from the View pull down menu.

5. Use the Place Constraint tool to place an Insert constraint with "0" offset between the circular edge of the hole on the Reel Spool and the inside circular edge of the Reel Handle Knob as shown in Figure 5.39. After selecting the circular edges, pick the Apply button to create the constraint.

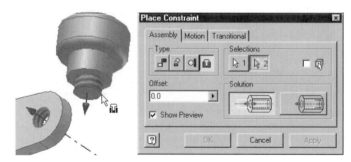

Figure 5.39

6. Use the Analyze Interference tool from the Tools menu and analyze the Reel Mounting Plate and the Reel Spool. After finding interference your screen will look similar to Figure 5.40.

Figure 5.40

7. Either edit in place or open the Reel Mounting Plate and change the diameter of Extrusion 3 to .4 inches. Save the changes and then reanalyze the Reel Mounting Plate and the Reel Spool for interference. There should be no interference between the two components.

8. Delete a few assembly constraints and then recreate them by using the Snap 'n Go Constraint (ALT drag) techniques.

9. When all the parts are constrained, save the changes to the assembly file and then close the file. The file will be used in a later tutorial.

TUTORIAL 5.4—TOP DOWN ASSEMBLY CONTINUED

In this tutorial you will apply assembly constraints, drive a constraint, and edit an assembly constraint.

1. Open the assembly as \Inventor Book\Linkage.iam.
2. Place an Insert constraint to align the holes and faces of the four components. Make sure the lower Long Link is grounded and the others are not grounded. When complete, your assembly should resemble Figure 5.41.

Figure 5.41

3. Test how the linkage will work by moving one of the short links or top long link. The assembled components should move honoring the assembly constraints.
4. Place an angle constraint between the two inside faces as shown in Figure 5.42. Enter an angle of −90 deg and then select OK to create the constraint.

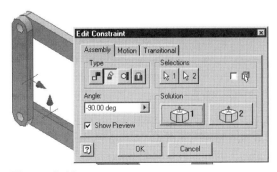

Figure 5.42

5. In the Browser right click on the angle constraint that you just created and select **Drive Constraint** from the menu. In the Drive Constraint dialog box change the End value to "0" and check Collision Detection as shown in Figure 5.43.

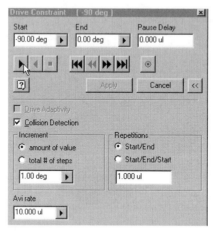

Figure 5.43

6. Pick the Forward button to see the motion. The motion will stop when the components collide. Pick the OK button to close the Collision Detected dialog box. At the top of the Drive Constraint dialog box, the angle of the collision will be noted. Then pick the Cancel button to return the components to their original position and stop the operation.

7. Edit the Insert constraints on the top of Short Links and for both select the front circular edge of the hole. Figure 5.44 shows the Edit Constraint dialog box with the number 2 button selected (depending upon the order that you selected the circular edges, yours may be number 1) and the front circular edge selected. After editing the first insert constraint, an alert box will appear warning you that a constraint is inconsistent. Accept the alert and then edit the other insert constraint on the top of the other Short Link. When done, your assembly should resemble Figure 5.45.

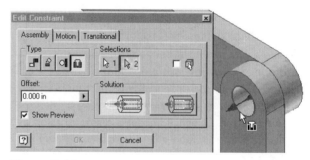

Figure 5.44

Assemblies 217

Figure 5.45

8. In the Browser right click on the angle constraint and select Drive Constraint from the menu. In the Drive Constraint dialog box change the End value to 90, check Collision Detection, change the incremental value to "2 deg" and change the Repetitions to Start/End/Start with a value of 4 ul as shown in Figure 5.46.

 Note: The ul stands for unitless.

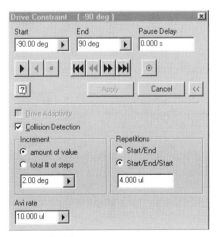

Figure 5.46

9. Select the Forward button to start the motion. The motion will go through four cycles (a cycle is 180 degrees) and components should not collide. When done pick the Cancel button to return the components to their original position and stop the operation.

10. Next you will create an animation (AVI) file. To reduce the file size decrease Inventor's graphics window to about half of full screen. (This will reduce the file size of the video).

11. In the Drive Constraint dialog box change the End value to 270, change the incremental value to 5 deg and change the Repetitions to Start/End with a value of 2 ul.

12. Pick the Record button; the Open dialog box will appear enter the following name: \Inventor Book\Linkage.avi and then pick the Open button. The Video Compression dialog box will appear and pick OK to accept the default compressor.

13. Select the Forward button to start the motion. The motion will go through two complete cycles. When done, pick the Record button to stop the recording. Then pick the Cancel button to return the components to their original position and stop the operation.

14. Play the \Inventor Book\Linkage.avi file that you just created. For comparison you can open the \Inventor Book\Linkage Example.avi that was included on the CD; your AVI file should resemble this one.

15. Maximize Inventor's screen size.

16. Save the assembly.

17. Close the assembly file.

iMATES

Another way to apply assembly constraints is to create iMates. An iMate holds information in the component or subassembly file on how the component or subassembly is to be assembled. An iMate needs to have the same name on both the components or subassemblies that are being constrained together. Each component or subassembly holds half of the iMate information and when put together they form a pair or a complete iMate. iMates are useful when like components are switched in an assembly. For example, you may have different pins that go into an assembly. The same iMate name can be assigned to each pin or a corresponding hole. When the pin is placed into the assembly, it will automatically be constrained to the corresponding hole. An iMate can only be used once in an assembly; once used it is consumed.

To create an iMate follow these guidelines.

- Pick the Create iMate tool on the Standard toolbar as shown in Figure 5.47.
- In the Create iMate dialog box, pick the type of constraint: Mate, Angle, Tangent, or Insert.
- In the graphics window, select the geometry you want to use as the primary position geometry.

- Click Apply. An iMate symbol will appear on the component and in the Browser as shown in Figure 5.48.
- Continue to create iMates, click Apply after each one.
- Save the file.

Figure 5.47   Figure 5.48

The symbol shows the type and state of the iMate. When an iMate is created, it is given a default name like: iInsert:1 or iAngle:1. The iMates can be renamed to better reflect the condition that they represent. Slowly double click on the iMate name in the Browser and type in a new name. In the Browser or on the iMate symbol in the graphics window, right click and select Properties from the context menu, then type in a new name.

To assemble components that exist in the same assembly and have corresponding iMates, there are two methods.

- Use the Place Constraint tool on the Assembly toolbar.
- Then pick an iMate symbol on a component then either select a matching iMate symbol on another component or drag the selected iMate symbol over a matching iMate symbol on another component. When the second iMate symbol is highlighted, click to position the components. Then pick the Apply button to create the constraint.

A component that has an iMate can automatically be constrained to another component that has the same iMate name.

- Issue the Place Component tool from the Assembly toolbar.
- Select the component to place and check the Use iMate check box in the lower left corner of the Open dialog box and click Open.
- If a matching iMate on another component exists, it will be placed and a consumed iMate symbol will be shown in the browser. Otherwise, repeat the sequence with the second component making sure to select Use Interface in the Open dialog box. The second component will be automatically constrained to the first.

Note: iMate names must match on the placed component and the unconsumed iMate in the assembly.

When the two matching iMates join in the assembly, a single consumed iMate is created. Because the relationship is specific to two components with matching iMate halves, multiple occurrences cannot be placed.

The solution that is selected for iMates (mate and flush, inside and outside, or opposed and aligned) must be the same for the matching iMates.

Multiple iMates can also be collected into a single set called a composite iMate. For more information on composite iMates see the online help.

REPLACING COMPONENTS

While designing, you may need to replace one component in an assembly with another component. A single component or all occurrences of the component can be replaced. The new component(s) will be placed in the same location if the origin of the replaced component is coincident with the origin of the placed component. The assembly constraint may be retained if the replaced component has the same shape as the original component. If the replaced component has a different shape than the original, some assembly constraints may need to be recreated to position the component correctly.

To replace a component follow these guidelines.

- In the Browser or the graphics window, select the component to replace; or on the Assembly toolbar, click the down arrow on the Replace Component tool, then choose either the Replace Component or Replace All tool as shown in Figure 5.49.

Figure 5.49

- In the Open dialog, select the component that will replace the existing component(s) and pick Open.

- A message notifies you that constraints and iMates will be retained, if possible. Pick OK to continue or Cancel to not replace a component.

PATTERNED COMPONENTS

When you are placing multiple occurrences of the same component or subassembly that will match a feature pattern on another part (component pattern), or will have a set of circular or rectangular patterns of a part in an assembly (assembly pattern), you can use the Pattern Component tool from the Assembly toolbar as shown in Figure 5.50. Both the component and assembly patterns will be described in the next section.

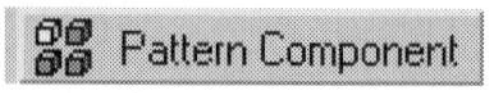

Figure 5.50

COMPONENT PATTERN

A component pattern will maintain a relationship to the feature pattern that is selected. For example, a bolt is component patterned to a bolt-hole pattern that consists of 4 holes. If the feature pattern (the bolt-hole) changes to 6 holes, the bolts will move to the new locations and two new bolts will be added for the two new holes. To create a component pattern, there must be a feature pattern and the part that will be patterned should be constrained to the parent feature (original feature that was patterned). Issue the Pattern Component tool from the Assembly toolbar, and the Pattern Component dialog will appear like what is shown in Figure 5.51. There are three tabs, Associative, Rectangular, and Circular. The Associative tab is the default and the tab that will be used to create a component pattern. By default the Component select option is active. Select the component or components that will be patterned and then pick the Feature Pattern Select button in the dialog box from where select a feature that is part of the feature pattern. Do not select the original feature. After selecting the pattern it will highlight on the part, and the pattern name will appear in the dialog box. When done, select OK to create the component pattern and exit the operation. The pattern can be edited by either selecting the pattern in the graphics window, or in the Browser, right click and pick Edit from the context menu. In the Browser, the component that was patterned will be consumed into a Component Pattern and the part occurrences will appear as elements, which can be expanded to see the part. You can suppress an individual part by right clicking on it and selecting Suppress from the context menu as shown in Figure 5.52. You can also break an individual part out of the pattern by right clicking on it and selecting Independent from the context menu. Once a part is independent, it has no relationship back to the pattern. To delete a pattern, either select the pattern in the graphics window, or in the Browser, right click and pick Delete from the context menu or pick the DELETE key from the keyboard. If the Delete Component Pattern Source(s) option from the Application Options, under the Assembly tab, is checked, the source component will automatically be deleted. If the source component should not be deleted when the pattern elements are deleted, you can uncheck this option.

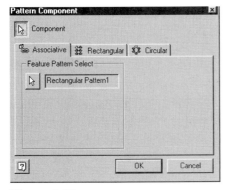

Figure 5.51

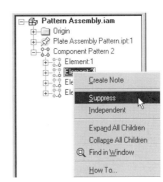

Figure 5.52

ASSEMBLY PATTERN

A component in an assembly can also be patterned either in a rectangular or in a circular fashion. The resulting pattern acts like a feature pattern. After creating it, you can edit it to change its numbers, spacing, etc. To pattern a component you must first create it or place it in an assembly. Then issue the Pattern Component tool from the Assembly toolbar and the Pattern Component dialog will appear. Pick the Rectangular or Circular tab and enter the placement values as needed. The options are the same that you learned about in the Pattern section in chapter three. The completed pattern then acts like a single part. If one part moves, they all move. The component that was patterned will be consumed into a Component Pattern in the Browser and each of the part occurrences will also appear as elements that can be expanded to see the part. The pattern can be edited by either selecting the pattern in the graphics window and right click, or right click on the pattern's name in the Browser and pick Edit from the context menu. At the part level, an individual part can be suppressed by right clicking on them and selecting Suppress from the context menu. Also at the part level you can break an individual part out of the pattern by right clicking on it and selecting Independent from the context menu. Once a part is independent, it has no relationship back to the pattern. To delete a pattern, either select the pattern in the graphics window, or in the Browser, right click and pick Delete from the context menu or pick the DELETE key from the keyboard. If the Delete Component Pattern Source(s) option from the Application Options, under the Assembly tab is checked, the source component will automatically be deleted. If the source component should not be deleted when the pattern elements are deleted, uncheck this option.

TUTORIAL 5.5—iMATES, REPLACING, AND PATTERN COMPONENTS

In this tutorial you will create three parts that have iMates. Assemble two of them, pattern the components, and then replace a component.

1. Start a new part file based on the default Standard.ipt file.
2. Create a 10 x 5 x 1 inch plate.
3. Place a .5 inch diameter through hole 1 inch up and over from the lower left corner.
4. Pattern the hole rectangular with 5 columns and 2 rows with 1 inch spacing between them.
5. Pick the Create Interface tool on the Standard toolbar and in the Create iMate dialog box pick the Insert type.
6. In the graphics area, select the first hole that you created (lower left); at this point your screen should resemble Figure 5.53, shown in an isometric view.

Then pick Apply to complete the operation, and an iMate symbol will appear on the component and in the Browser. Click Cancel to close the Create iMate dialog box.

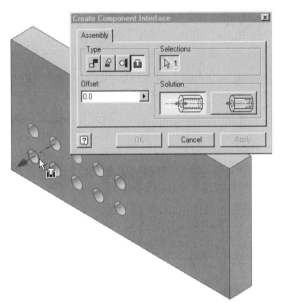

Figure 5.53

7. Select the iMate that you just created in the Broswer, below the iMates folder, or in the graphics window and right click and select Properties from the context menu. Change the Name to Pin Insert.
8. Save the part as \Inventor Book\Plate Pattern.ipt.
9. Close the file.
10. Start a new part file based on the default Standard.ipt file.
11. Create a .5 inch diameter cylinder that is 2 inches long.
12. Change to an isometric view.
13. Make a new sketch on the back circular edge.
14. Draw a .75 inch diameter circle that is concentric to the circular edge.
15. Extrude both the projected .5 and .75 inch diameter circles .125 inches away from the part, joining material.
16. Pick the Create iMate tool on the Standard toolbar and in the Create iMate dialog box pick the Insert type.

17. In the graphics area, select the inside .75 inch circular edge as shown in Figure 5.54 and then pick Apply. An iMate symbol will appear on the component and in the Browser. Click Cancel to close the Create iMate dialog box.

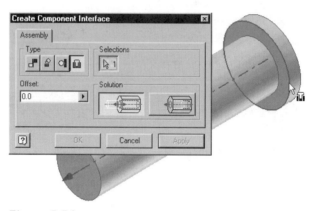

Figure 5.54

18. Select the iMate that you just created in the Browser, below the iMates folder, or in the graphics window and right click and select Properties from the context menu. Change the Name to Pin Insert.
19. Save the part as \Inventor Book\Pin Three Quarter Flange.ipt.
20. From the File pull down menu select Save Copy As.
21. Save the copied part as \Inventor Book\Pin One Inch Flange.ipt.
22. Close the file.
23. Open the file \Inventor Book\Pin One Inch Flange.ipt.
24. Change the .75 inch diameter flange (Extrusion2) to 1 inch.
25. Update and save the part.
26. Close the file.
27. Start a new assembly file based on the default Standard.iam file.
28. Place one occurrence of the part \Inventor Book\Plate Pattern.ipt into the assembly.
29. Place one occurrence of the part \Inventor Book\Pin Three Quarter .ipt into the assembly. Before picking Open, check the Use iMate option in the Open dialog box. Then pick Open and the pin will automatically be constrained to the hole.
30. Start the Replace Component tool from the Assembly toolbar. Select the pin as the part to be replaced and then from the Open dialog box pick the part

\Inventor Book\Pin One Inch Flange.ipt and click Open. The constraints will be maintained because they have the same iMate name.

31. Select the Pattern Component tool from the Assembly tool bar and then select the Pin as the component and then select the rectangular pattern using the Feature Pattern Select button. After making the selections, your screen should resemble Figure 5.55. Click OK to create the pattern.

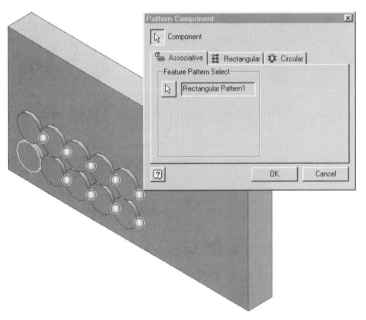

Figure 5.55

32. Change the number of holes and spacing between the holes on the plate by editing the plate. While still in the assembly, either double click on the plate's name in the Browser or double click on the plate itself in the graphics area. You can also right click on the plate's name in the Browser and pick Edit from the context menu.

33. Edit the Rectangular Pattern to 2 columns with 8-inch spacing and 2 rows with 3-inch spacing as shown in Figure 5.56. Click OK to close the Rectangular Pattern dialog box.

34. Update the part and return to the top-level assembly by either double clicking on the Assembly's name in the Browser or selecting Return from the Command bar. When done, your screen should resemble Figure 5.57.

35. Close the file without saving changes.

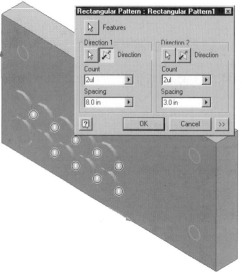

Figure 5.56   Figure 5.57

ASSEMBLY WORK FEATURES

When placing assembly constraints, you may need to create a constraint in a location that does not lie on a part. Work planes, work axis, and work points can be created in the context of an assembly and between two parts. While in an assembly, use the work plane, work axis, or the work point tool from the Assembly Panel Bar and create the work feature like you would in a part. Work features created in the context of the assembly only exist in the assembly and are associated to the parts face, edge, etc. that were selected.

ADAPTIVITY

Adaptivity is the functionality in Inventor that allows the size of a part to be determined by setting a relationship between the part and another part in the assembly. Adaptivity allows under-constrained sketches; features that have undefined angles or extents; hole features; and subassemblies, which contain parts that have adaptive sketches or features, to be adaptive. The adaptivity relationship is acquired by applying assembly constraints between an adaptive sketch or feature and another part. If a sketch is fully constrained, it cannot be made adaptive, but the extruded length or revolved angle can be. A part can only be adaptive in one assembly at a time. In an assembly that has multiple placements of the same part, only one occurrence can be adaptive. The other occurrences will reflect the size of the adaptive part. An example of adaptivity would be to have the diameter of a pin get its size from a hole. In the same example you could have the hole get its diameter from the pin. Adaptivity can be turned on and off as needed. Once a part's size is determined through adaptivity, you may want to turn its adaptivity off. If you want to create adaptive features, there are options within Inventor that will speed the process of creating them. From the Tools pull down menu, pick Application Options; from the Assembly tab there are three areas that relate to adaptivity as shown in Figure 5.58. Each option will be described.

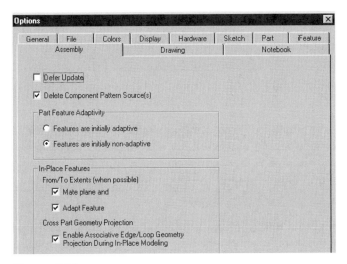

Figure 5.58

Part Feature Adaptivity: Here you will select if you want features to be adaptive or non-adaptive when they are created.

> **Features Are Initially Adaptive:** When checked, features will be adaptive when they are created. Unless you are creating many adaptive features, this option should not be selected.

> **Features Are Initially Non-adaptive:** When checked, features will not be adaptive when they are created. Unless you are creating many adaptive features, this option should be selected.

In-Place Features From/To Extent (when possible): Here you will determine if a feature will be adaptive when the To or From/To option is selected for the Extent. If both options are selected, Inventor will try to make the feature adaptive, but if it cannot, it will terminate at the selected face.

> **Mate Plane:** Select this option when you want to create a new component and have a mate constraint applied to the plane on which it was constructed, but it will *not* be adaptive.

> **Adapt Feature:** Select this option when you want to create a new component and have it adapt to the plane on which it was constructed.

Cross Part Geometry Projection

Enable Associative Edge/Loop Geometry Projection During In-Place Modeling: When checked and geometry is projected from another part onto the active sketch, the projected geometry is associative (sketch associativity) and will update when changes are made to the parent part. Projected geometry can be used to create a sketched feature.

To make a sketch or a feature adaptive, it must exist in the assembly that will be adaptive.

Following are some guidelines on making sketches, features, and subassemblies adaptive.

SKETCHES

- Draw the profile and do *not* place parametric dimensions to the areas that you want to become adaptive. If you want to place a dimension to the area that will be adaptive, use a driven dimension.

- Right click on the sketches name in the Browser and from the context menu pick Adaptive as shown in Figure 5.59.

- When done, a small glyph (two circular arcs with arrows) will appear to the left of the sketch's name in the Browser as shown in Figure 5.60. Once the sketch is turned into a feature, the feature will also be adaptive.

Figure 5.59

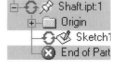

Figure 5.60

Or

- Draw the profile and do *not* place parametric dimensions to the areas that you want to become adaptive. If you want to place a dimension to the area that will be adaptive, use a driven dimension.

- Extrude, revolve, etc. the profile, then right click on the new feature in the Browser and from the context menu pick either Adaptive or Properties. If you select Adaptive, both the sketch and feature will become adaptive. If you select Properties, the Feature Properties dialog box will appear as shown in Figure 5.61. Here you can make the Sketch, Parameters, or From/To Planes adaptive.
- When done, a small glyph (two circular arcs with arrows) will appear to the left of the sketch and feature's name in the Browser.

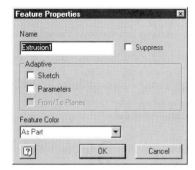

Figure 5.61

FEATURES

- Right click on the features name in the Browser and from the context menu Select Adaptive or Properties. If you select Adaptive, both the sketch and feature will become adaptive. If you select Properties, a Feature Properties dialog box will appear as shown back in Figure 5.61. Here you can make the Sketch, Parameters, or From/To Planes adaptive.

Or

- Before the feature is created, pick Application Options from the Tools pull down menu and from the Assembly tab check the option Features are initially adaptive. Then create the feature and it will be adaptive.

SUBASSEMBLY

In an assembly, you can also make subassemblies adaptive. At the top assembly level right click on the subassembly's name and check Adaptive from the context menu. For features within a subassembly to adapt, the feature, part, and subassembly name must all be adaptive.

 Note: To turn off the adaptivity for sketches, features, and subassemblies, uncheck Adaptivity from the context menu or from the Feature Properties dialog box.

ADAPTING THE SKETCH OR FEATURE

After making a sketch or feature adaptive, the part itself must be made adaptive. To make a part adaptive, make the top level of the assembly where the part exists active. Then right click on the parts name and select Adaptive from the context menu. Apply assembly constraints that will define the relationship for the adaptive sketch or feature. As the assembly constraints are being applied, degrees of freedom are being removed.

Note: Parts that are imported from a SAT or STEP file format cannot be made adaptive because they are static and do not have sketches and features.

In assemblies that have multiple adaptive parts, two Updates may be required to solve correctly.

For revolved features use only one tangency constraint.

Avoid offsets when applying constraints between two points, two lines, or a point and a line. Incorrect results may occur.

TUTORIAL 5.6—ADAPTIVITY

In this tutorial you will make a sketch and feature adaptive, and then reverse the adaptivity between the parts

1. Start a new assembly file based on the default Standard.iam file.
2. Create a new component in the assembly file name Plate Adapt.
3. Sketch and dimension a 2 inch square and then extrude it 1 inch.
4. Change to an isometric view.
5. Create a .5 inch diameter hole that is centered in the middle of the part that uses the Through All termination. When done your screen should resemble Figure 5.62.

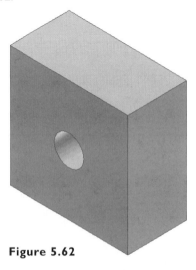

Figure 5.62

Tip: Create a hole center to place the hole.

6. Make the top level assembly active by either picking the Return tool from the Command Bar or double click on the top level assembly's name.

7. Create a new component in the assembly file name Pin Adapt and make sure the option Constrain sketch plane to selected face or plane is checked. Then select the front left vertical face with the hole through it as the face to place the active sketch on.

8. From the Tools pull down menu pick Application Options and from the Assembly tab confirm that the Mate Plane and Adapt Features options are selected.

9. Draw a small circle near the center of the hole.

10. Make the sketch Adaptive.

11. Extrude the circle using the To option for the Extents and select the back of the plate as the termination face as shown in Figure 5.63.

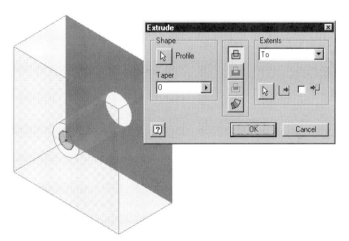

Figure 5.63

12. Make the top level assembly active.

13. Apply a Mate assembly constraint that has an Offset value of .06. Select the inside circular face of the hole and the outside circular face of the pin and as shown in Figure 5.64. When applied, the pin's diameter should be .06 inches in from the hole. Close the Place Constraint dialog box.

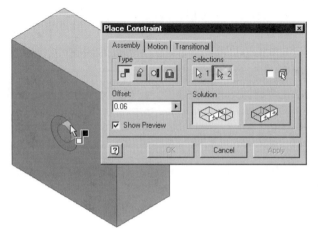

Figure 5.64

14. Make the part "Plate Adapt" active.
15. Edit the plate and change the extrusion distance to 3 inches and the hole diameter to 1 inch. Update the part and make the top level assembly active. When done, your screen should resemble Figure 5.65.

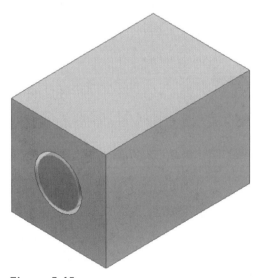

Figure 5.65

16. Now you will reverse the adaptivity and make the hole diameter adapt to the pin's diameter.

17. At the assembly level remove the adaptivity from the part "Pin Adapt" by right clicking on its name in the Browser and uncheck Adaptive from the context menu.
18. Make the part "Pin Adapt" active.
19. Edit the sketch and place a 1.5 inch diameter dimension on the circle.
20. Pick the Return tool on the Command Bar to exit the Sketch environment and return to modeling environment.
21. Make the top level assembly active. You will get an error message alerting you that the pin is larger than the hole and the Mate constraint is inconsistent. Click Accept to dismiss the error message. In the Browser a yellow circle with an exclamation point will appear by the Mate constraint alerting you to the inconsistency.
22. Make the part "Plate Adapt" adaptive and again accept the error message.
23. Make the part "Plate Adapt" active and again accept the error message.
24. If needed, expand the tree for the active part and right click on the Hole feature and check Adaptive.
25. Pick the Return tool on the Command Bar to exit modeling and return to the base assembly. When done your screen should resemble Figure 5.66.
26. Practice changing the values of the constraints and parts adaptivity.
27. When done, close the file without saving the changes.

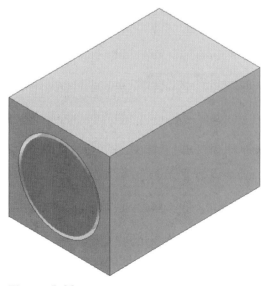

Figure 5.66

ENABLED COMPONENTS

When working in an assembly, you may want to hide a part so that other parts can be seen. The visibility of parts can be turned on or off by selecting its name in the Browser or on the part itself and right clicking and selecting Visibility from the context menu. When the visibility of a part is off in an assembly, that part can still be used in other assemblies and can be opened and edited. Inventor gives you another option for controlling how parts look in an assembly called Enabled. By default, all parts are enabled when they are placed or created; their visibility is on, and they are displayed in the current display mode. When a part is enabled, it is slowing down the graphics regeneration speed because it has to be calculated each time a view is dynamically rotated or panned. A part can be disabled. When a part is disabled, it is displayed as a wireframe and cannot be selected in the graphics area. However, a disabled part can have geometry projected from it. To disable a component, select its name in the Browser or on the part itself and right click and pick Enabled from the context menu as shown in Figure 5.67. Figure 5.67 also shows how the spool of the reel looks when it is disabled.

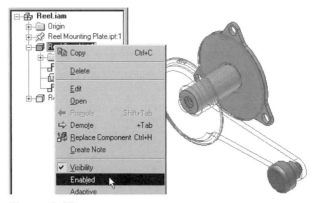

Figure 5.67

DESIGN VIEWS

While working in an assembly you may want to save configurations that show the assembly in different states and from different view points. Design views can store the following information: component visibility (visible or not visible), component selection status (enabled or not enabled), color settings and style characteristics applied in the assembly, zoom magnification, and viewing angle. The information in the design view is saved to an associated file that has the file extension of idv. It is recommended to save the Design View file with the same name and in the same directory as the assembly file. There can be multiple Design Views saved in the same idv file. Design views can be used while working on the assembly and can be used when creating presentation views or drawing views. Once the screen orientation and part visibility is set, you can create a Design View by picking the arrow next to the Design Views icon from the top of the Browser and pick Other as shown in Figure 5.68, or pick Design Views from the View pull down menu. The Design Views dialog will appear like what is shown in Figure 5.69. Type in or select a file name and then type in a name for the new Design View and then pick the Save button. To make a Design View current, either select it from the drop list where you picked the Other options or after picking Other, select its name from the list and then pick the Apply button. To delete a Design View, once in the Design View dialog box, select the Design View's name from the list and then pick the Delete button.

Figure 5.68

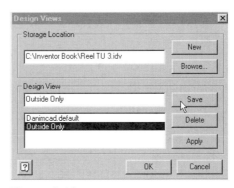

Figure 5.69

PRESENTATION FILES

After creating an assembly, you can create drawing views based on how the parts are assembled. If you need to show the components in different positions like an exploded view, hide specific components or create an animation that shows how to assemble or disassemble the components you will need to create a presentation file. A presentation file is a separate file from an assembly and has a file extension of ipn. The presentation file is associated to an assembly file; changes made to the assembly file will be reflected in the presentation file. No components are created in a presentation file. To create a new presentation file, select the New icon from the What To Do section and then select the Standard.ipn icon from the Default template tab as shown in Figure 5.70. Before we start creating presentation files, let's look at the presentation tools that are available. By default the panel bar will show the Presentation tools as shown in Figure 5.71 with the expert mode turned off (text descriptions shown).

- **Create View:** Use this tool to create views of the assembly. There is no limit to the number of views that can exist in a presentation file, but all the views are based on the same assembly. Once created, a view will be listed as an Explosion in the Browser.

- **Tweak Components:** Use this tool to move and/or rotate parts in the view.
- **Precise View Rotation:** You can rotate the view by a specified angle direction using a dialog box.
- **Animation:** Use this tool to create an animation of the assembly; an AVI file can be output.

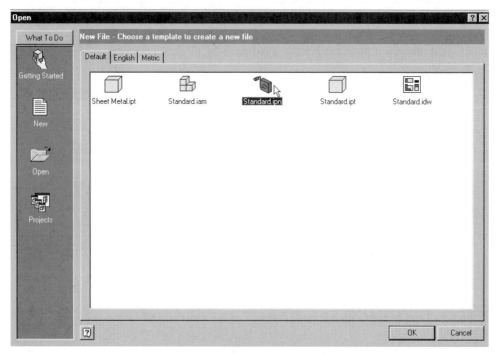

Figure 5.70

Figure 5.71

The main steps for creating a presentation view are to reposition (tweak) the parts in specific directions and then create an animation if needed. Each of the steps will be discussed in the following sections.

CREATING PRESENTATION VIEWS

The first step in creating a presentation is to create a presentation view. Start the Create View tool from the Presentation Panel Bar or right click and pick Create View from the context menu. The Select Assembly dialog box will appear as shown in Figure 5.72. The dialog box is divided into two sections, Assembly and Explosion method, each will be described in this section.

Figure 5.72

Assembly: In this section you will determine what assembly and design view the presentation view will be based on.

> **File:** Enter or select an assembly file that the presentation view will be based on.
>
> **Design View:** In this area the active design view will be displayed. If you want to use a different design view, enter its name or select it.

Explosion Method: In this area you can choose to manually or automatically have the parts be exploded (separate the parts a given distance).

> **Manual:** If selected, the components will not automatically be exploded. After the presentation view is created, tweaks can be added that will move or rotate the parts.
>
> **Automatic:** Select Automatic if you want the part to be exploded a given value.
>
> **Create Trails:** If Automatic is selected, you can choose to have visible trails (lines that show how the parts are being exploded).
>
> **Distance:** If Automatic is selected, enter a value that the parts will be exploded.

After making your selection, pick the OK button to create a presentation view. The components will appear in the graphics window and a presentation view will appear in the Browser. If the automatic option was checked for the Explosion Method, the parts will automatically be exploded. To determine how the parts will be exploded, Inventor analyzes the assembly constraints. If parts were constrained using the mate plane option and the normals (arrows perpendicular to the plane) for both parts were pointing outward, they will be exploded away from each other. Think of the mating parts as two magnets that want to push themselves in opposite direction. The grounded component in the Browser will be the component that stays stationary, and the other components will move away from it. If trails were automatically created, they will generally come from the center of the part not necessarily from the center of the holes. After expanding all the children in the Browser, you can see numerically how the parts exploded. Once the parts are exploded, the distance is referred to as a tweak. The number by each tweak reflects the distance that the component is moved from the base component.

TWEAK COMPONENTS

After creating the presentation view, you may choose to edit the tweak of a component or move or rotate the component beyond the original explosion factor. To edit an individual tweak, in the Browser double click on its value as shown in Figure 5.73. Then type in a new value in the text box in the lower left corner of the Browser and press ENTER to have the new value be used.

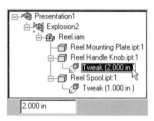

Figure 5.73

To extend all the tweaks the same distance, in the Browser right click on the assembly's name and select Auto Explode from the context menu as shown in Figure 5.74. Then enter in a value and then select the OK button and all the tweaks will be extended the same distance. Negative values cannot be used with the Auto Explode method.

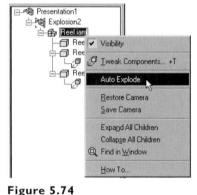

Figure 5.74

Another method to manually reposition the components is to tweak them using the Tweak tool. To manually tweak a component(s), start the Tweak Component tool from the Presentation Panel bar, press the hot key T or right click and pick Tweak Component from the context menu. The Tweak Component dialog box will appear as shown in Figure 5.75. The Tweak Component dialog box is broken into two sections, Create Tweak and Transformations. Each section will be described next.

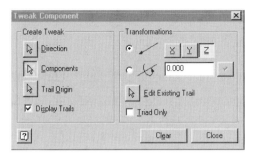

Figure 5.75

Create Tweak
In this section you will select the components to tweak and set the direction and origin of the tweak.

Direction: Here you will determine the direction or axis of rotation for the tweak. After picking the Direction button select an edge, face, or feature of any component in the graphics window to set the direction triad (x, y and z) for the tweak. The edge, face, or feature that is selected does not need to be on the components that are being tweaked.

Components: Here you will select the components to tweak. Click the Components button and then select the components in the graphics window or Browser that will be tweaked. If a component was selected when you started the Create Tweak operation, it will automatically be included in the components. To remove a component from the group, press and hold the CTRL key and select the part.

Trail Origin: Here you will set the origin for the trail. Click the Trail Origin button, then click in the graphics window to set the origin point. If you do not specify the trail origin, it will be placed at the center of mass for the part.

Display Trails: Check the Display Trails box if you want to see the tweak trails for the selected components; clear the check box to hide the trails.

Transformations
In this section you will set the distance and type of tweak.

Linear: To move the component in a linear fashion, pick the button next to the arrow and line.

Rotation: To rotate the component, pick the button next to the arrow and arc.

X, Y, Z: Pick the X, Y, or Z coordinate button to determine the direction for a linear tweak or the axis for a rotational tweak. Or select the arrow on the triad that represents the X, Y, or Z direction.

Text Box: Enter a positive or negative value for the tweak distance or rotation angle. Or pick a point in the graphics area and with the mouse button pressed down move the mouse to set the distance.

Apply: After making all the selections, pick the Apply button to complete the tweak.

Edit Existing Trail: To edit an existing tweak, pick the Edit Existing Trail button and then select the tweak in the graphics window and change the desired settings.

Triad Only: To rotate the direction triad without rotating selected components, select the Triad Only option, enter the angle of rotation, and then click the Apply button. After the triad is rotated, you can use it to define tweaks.

Clear: Pick the Clear button in the dialog box to remove all the settings and set up another tweak.

To tweak a component follow these guidelines.

- Issue the Tweak Component tool.
- Determine the direction or axis of rotation for the tweak by picking the Direction button and then select an edge, face, or feature.
- Select the components to tweak by picking the Components button and then select the components in the graphics window or Browser that will be tweaked.
- Select a trail origin point.
- Determine if you want trails visible or not.
- Set the type of tweak to linear or rotation.
- Pick the X, Y, or Z coordinate button to determine the direction for a linear tweak or the axis for a rotational tweak.
- Enter a value for the tweak in the text box, in the Tweak Component dialog box, or pick a point on the screen and drag the part into its new position.
- Pick the Apply button in the Tweak Component dialog box.

ANIMATION

After the components have been repositioned, you can animate the components showing how they assemble or disassemble. To create an animation, pick the Animate tool from the Presentation Panel bar or right click and pick Animate from the context menu. The Animation dialog box will appear as shown in Figure 5.76. The Animation dialog box is broken into three sections Parameters, Motion, and Animation Sequence. Each section will be described next.

Parameters

In this section you will specify the number of repetitions for the animation.

Assemblies 241

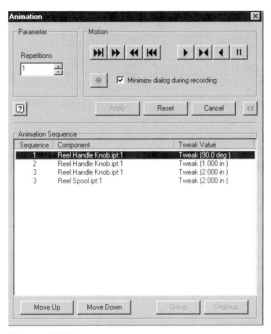

Figure 5.76

Repetitions: Here you will set the number of times to repeat the playback. Enter the desired number of repetitions or use the up or down arrow to select the number.

 Note: To change the number of Repetitions, click the Reset button on the dialog box and then type in a new value.

Motion
In this section you will play the animation for the active view.

- **Forward by Tweak:** Drives the animation forward one tweak at a time.
- **Forward by Interval:** Drives the animation forward one interval at a time.
- **Reverse by Interval:** Drives the animation in reverse one interval at a time.
- **Reverse by Tweak:** Drives the animation in reverse one tweak at a time.
- **Play Forward:** Plays the animation forward for the specified number of repetitions. Before each repetition, the view is set back to its starting position.
- **Auto Reverse:** Plays the animation for the specified number of repetitions. Each repetition plays start to finish, then in reverse.
- **Play Reverse:** Plays the animation in reverse for the specified number of repetitions. Before each repetition, the view is set back to its ending position.

- **Pause:** Pauses the animation playback.
- **Record:** Records the specified animation to a file so that you can play it back later.

Animation Sequence

In this section you can change the sequence in which the tweaks happen, select the tweak, and then pick the needed operation.

Move Up: Moves the selected tweak up one place in the list.

Move Down: Moves the selected tweak down one place in the list.

Group: Select a number of tweaks and then select the Group button. When tweaks are grouped, all the tweaks in the group will move together as you change the sequence. When tweaks are grouped, the group assumes the sequence order of the lowest tweak number.

Ungroup: After selecting a tweak that belongs to a group, you can pick the Ungroup button, and it can then be moved individually in the list. The first tweak in the group assumes a number one-higher than the group. The remaining tweaks are numbered sequentially following the first.

To animate components follow these guidelines.

- Issue the Animate tool.
- Set the number of repetitions.
- Adjust the tweaks as needed.
- Click one of the play buttons to view the animation in the graphics window.
- To record the animation to a file, click the record button, then click one of the play buttons to start recording.

TUTORIAL 5.7—PRESENTATION VIEWS

In this tutorial you will create a presentation view, add tweaks, and create an animation. In tutorial 5.8 you will use the presentation view for the drawing view.

1. Start a new presentation file based on the default Standard.ipn file.
2. Start the Create View tool.
3. Select the file \Inventor Book\Reel.iam, leave the Design View as the default, change the Explosion Method to Automatic and change the distance to 2 as shown in Figure 5.77, and then pick the OK button to create a presentation view.

Assemblies 243

Figure 5.77

4. Start the Tweak Components tool.
5. Set the direction by picking the back of the Reel Mounting Plate, select the Reel Handle Knob as the component to tweak, select X as the linear direction, and enter a distance of 1 shown in Figure 5.78. Then pick the Apply button. Click Clear to begin the definition of another Tweak.

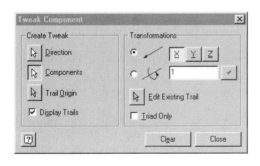 

Figure 5.78

6. Set the direction by picking the front of the Reel Handle Knob, select the Reel Handle Knob as the component, select the Rotation button, select X as the rotation axis and enter 90 degrees as shown in Figure 5.79. Then pick the Apply button. Click Close to close the Tweak Component dialog box. When done your components should look like the ones shown in Figure 5.80.

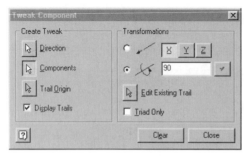

Figure 5.79

7. Start the Animate tool.

8. Change the Repetitions to 2 and click the Apply button; select the Record button, type in a file name \Inventor Book\Reel Explode.avi. Select *save* and the Video Compression dialog box will appear, pick OK to accept the default compressor and then select the Auto Reverse button to create an animation file as shown in Figure 5.80. After completing the animation, you can compare your file against \Inventor Book\Reel Explosion Example.avi that was included on the CD.

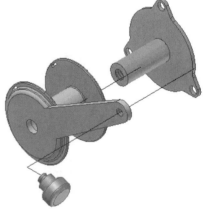

Figure 5.80

9. While still in the same file, start the Create View tool.

10. You will now recreate the same explosion using the manual process. In the Select Assembly dialog box, change the Explosion Method to Manual and then pick the OK button to create another presentation view.

11. Start the Tweak Components tool.

12. Set the direction by picking the front of the Reel Spool, select the Reel Spool, and Reel Handle Knob as the components to tweak; select Z as the linear direction, enter a distance of 2; then pick the Trail Origin tool and select a point near the center of the hole of the Reel Spool shown in Figure 5.81 and then pick the Apply button. Click Clear to start the definition of another tweak.

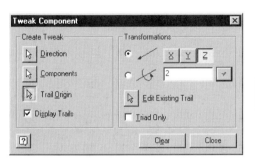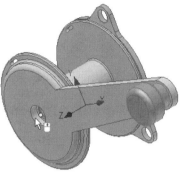

Figure 5.81

13. Try to duplicate the previous results. You will need to tweak the Reel Handle Knob out away from the Reel Spool an additional 2 inches and then follow the same steps that you did in steps 5 and 6.

14. Then create an animation to check your results.

15. When done save the presentation file as \Inventor Book\Reel.ipn.

16. Close the file.

CREATING DRAWING VIEWS FROM ASSEMBLIES AND PRESENTATION FILES

After creating an assembly, you may want to create drawings that utilize the data. Drawing views from assemblies can be created using information from an assembly file, or a presentation file and from them you can pick a specific design view or presentation view. To create drawing views based on assembly data, start a new drawing file or open an existing drawing file like you learned in chapter 4. Then use the Create View tool and select the IAM or IPN file to create the drawing from. If needed, specify the design view or presentation view in the dialog box. Figure 5.82 shows a presentation view being picked after a presentation file was selected. The drawing can contain as many drawing views based on different IAM and IPN files as needed. The same rules apply to creating drawing views from assemblies or presentation files as part (IPT) files.

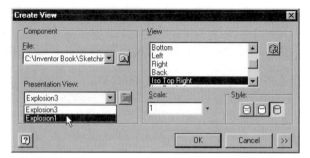

Figure 5.82

CREATING BALLOONS

After the drawing view(s) are created, you can add balloons to the parts. Before ballooning your drawing, make changes to the balloon style by picking the Standards tool from the Format pull down menu. From the Balloon tab pick the desired style. These changes will only be used in this drawing file. Save the changes to a template file if new drawings will use the same style.

To add balloons to a drawing follow these guidelines.

- Activate the Drawing Annotation Panel Bar by picking in the panel bars title on Drawing Management and select Drawing Annotation from the context menu.

- From the Drawing Annotation Panel Bar, pick the arrow next to the Balloon tool as shown in Figure 5.83 and two icons will appear: the first for ballooning components one at a time and the second for ballooning all the components in a view in a single operation. The hot key B can also be used to create balloons for single components.

- To balloon a single component, pick the Balloon tool from the Drawing Annotation Panel Bar or use the hot key B.

- Select a component to balloon.

- The Parts List Item Numbering dialog box will appear as shown in Figure 5.84.

- From within the Parts List Item Numbering dialog box, choose what Level, Range, and Format to use. Descriptions of the options available in the dialog box are given in the next section.

 Note: The same dialog box will be used to create a parts list.

- A preview image of the balloon will appear.

- Position the mouse to place the second point for the leader and then press the left mouse button.

- Continue to select points to add segments to the balloon's leader.
- When done adding segments to the leader, right click and from the context menu select Continue to create the balloon.
- Pick the Done option to cancel the operation without creating the balloon or select the Backup option to undo the last step that was created for the balloon.

Figure 5.83

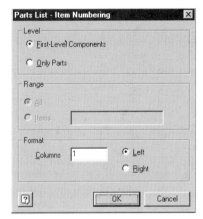

Figure 5.84

- To balloon all the components in a view, pick the second icon (Balloon All) from the Balloon tool.
- Pick a point in the view that the balloons will be created in.
- The Parts List Item Numbering dialog box will appear as shown in Figure 5.84.
- From within the Parts List Item Numbering dialog box, choose what Level, Range and Format to use. Descriptions of the options available in the dialog box are given in the next section.

 Note: The same dialog box will be used to create a parts list.

- After making your selection, pick the OK button and the balloons will be created in the selected view.

PARTS LIST ITEM NUMBERING DIALOG BOX OPTIONS

Level: Specify what level should be included when ballooning components.

First-Level Components: Pick this option if you want to only balloon the top-level components of the assembly in the selected view. Subassemblies and parts that are not part of a subassembly will be ballooned, but parts that are in a subassembly will not be.

Only Parts: Pick this option to balloon only the parts of the assembly in the selected view. Subassemblies will not be ballooned, but the parts in the subassemblies will be ballooned.

Range: This section is not used for creating balloons, only for creating parts lists.

All: Creates a parts list for all the components in the selected view.

Items: Creates a parts list for the specified range of parts. Select Items then enter the part numbers, separated by commas, or click the parts in the graphics window to include them in the list.

Format: Specifies the number of columns to split the parts list into. Splitting parts lists into multiple columns is useful when parts lists are long and do not fit on the drawing sheet as desired.

Columns: Creates a parts list with the specified number of columns. Type in the number of columns you want to split the parts list into.

Left: Attaches the cursor to the top right corner of the parts list while placing the parts list on the drawing sheet. This allows you to use the top right corner of the parts list as an insertion point for the parts list. You will want to use this option if placing the Parts List on the right side of the drawing sheet.

Right: Attaches the cursor to the top left corner of the parts list while placing the parts list on the drawing sheet. This allows you to use the top left corner of the parts list as an insertion point for the parts list. You will want to use this option if placing the Parts List on the left side of the drawing sheet.

To edit a balloon after it has been created, move the mouse over the balloon and a small green dot will appear at its vertices. Right click on one of the green dots to access a context menu. You can delete, change the arrowhead's appearance, attach another balloon to it, remove an attached balloon, or add a segment or vertex.

 Note: A parts list can be created without creating balloons.

PARTS LIST

After the drawing views are created, you can also create a parts list. As noted above, components do not need to be ballooned before creating a parts list. To create a parts list, pick the Parts List tool from the Drawing Annotation toolbar. Select a view that the parts list will be based on. If balloons were not created, the Parts List Item Numbering dialog box will appear as was discussed in the ballooning section. After specifying your options, pick the OK button and a parts list will appear. If the components were already ballooned, the parts list will appear without first seeing the Parts List Item Numbering dialog box. Pick a point to locate the parts list. The information in the parts list is populated from the properties of each component. A long parts list can also be split into multiple columns. Follow these guidelines to split a parts list that has already been placed on the drawing.

- Right click on the parts list and pick Edit Parts List from the context menu.
- From the Edit Parts List dialog box, pick the row that the split will occur after.
- Right click and pick Column Split from the context menu as shown in Figure 5.85.
- To complete the operation Pick the OK button.

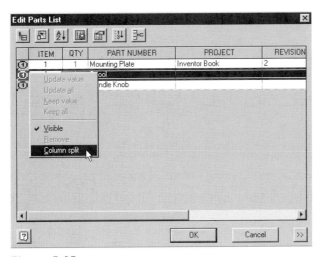

Figure 5.85

TUTORIAL 5.8—ASSEMBLY DRAWING VIEWS

In this tutorial, you will create drawing views based on an assembly and a presentation file, balloon the components, and create a parts list.

1. Start a new drawing file based on the default Standard.idw file.
2. Change the border to a "B" size sheet.
3. Pick the Create View tool from the Drawing Management Panel Bar.
4. Select the File \Inventor Book\Reel.iam, leave the scale at 1 and use the front view and place the view in the lower left corner of the border.
5. Then project a top, right, and isometric view. When done, your screen should resemble Figure 5.86.

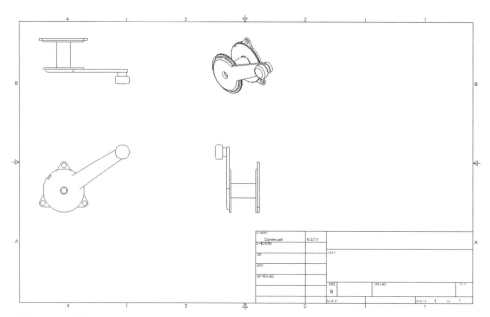

Figure 5.86

6. Pick the Create View tool from the Drawing Management Panel Bar.

7. Select the File \Inventor Book\Reel.ipn, leave the scale at 1 and change the Presentation View to Explosion1; then change the type of view to Iso Top Right and place the view on the right side of the border.

8. Edit the explosion view and change the Style to shaded.

9. Change to the Drawing Annotation tools in the Panel Bar.

10. Use the Balloon All tool to balloon all the components in the presentation view. When done, your screen should resemble Figure 5.87.

 Note: The component numbers may be different.

11. Open the file Reel Mounting Plate and edit its Properties. Under the Project tab change the Revision to 2, and the Project to Inventor Book as shown in Figure 5.88 and from the Summary tab change the Author to Your Name. Click OK to close the Properties dialog box and apply the changes.

12. Save the part file and then close it.

13. Back in the drawing file create a parts list based on the exploded view. Place it on top and to the right of the title block.

Assemblies 251

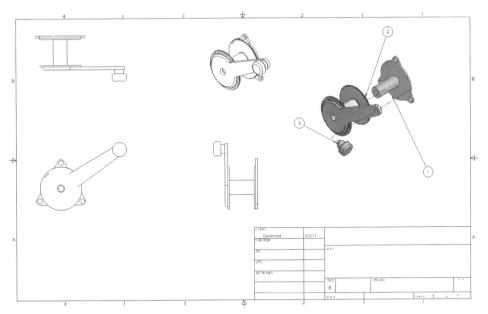

Figure 5.87

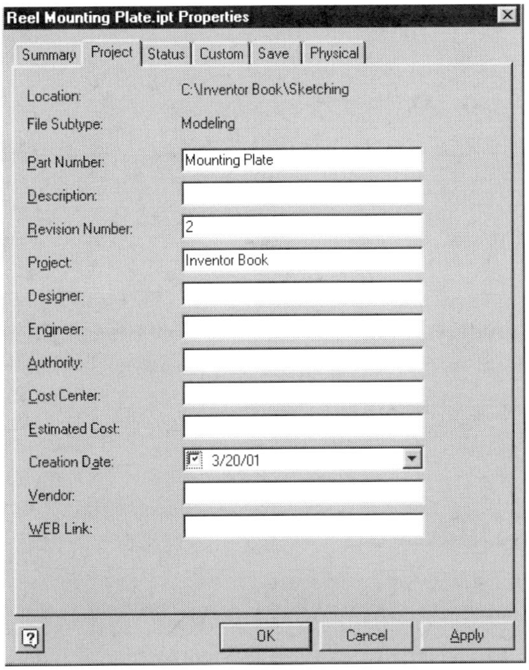

Figure 5.88

14. Edit the parts list and use the Column Chooser tool to delete the Description property and add the Properties; Project and Revision Number to the Selected Properties column as shown in Figure 5.89. When done, your parts list should resemble Figure 5.90.

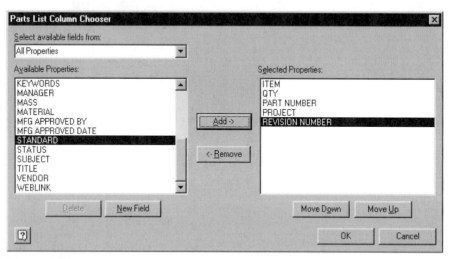

Figure 5.89

Parts List				
ITEM	QTY	PART NUMBER	PROJECT	REVISION NUMBER
1	1	Mounting Plate	Inventor Book	2
2	1	Spool		
3	1	Handle Knob		

Figure 5.90

15. As time permits, edit the width and names of the columns and add missing annotations.
16. When done, save the file as \Inventor Book\Reel.idw.
17. Close the file.

PRACTICE EXERCISES

Exercise 5.1—Adaptive Assembly

In this exercise you will use adaptivity to solve the length of a cylinder. You will place two components and create another part in the assembly.

1. Start a new assembly file based on the default Standard.iam file.
2. Place the Component Hydraulic Base and Hydraulic Press. Only the Hydraulic Base should be grounded.
3. Constrain the Hydraulic Base and Hydraulic Press using an Insert constraint. When done, rotate the assembly so your screen resembles Figure 5.91.

Figure 5.91

4. Create a new component named Hydraulic Cylinder. Do not constrain it to another part but position the sketch to one of the front vertical faces. Sketch an oval shape and dimension the radius with a 1 inch radius and then make the sketch *adaptive* as shown in Figure 5.92. Extrude the oval 1 inch and add two 1-inch diameter holes to the center of both circular edges.

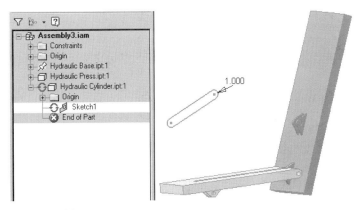

Figure 5.92

5. Constrain the cylinder to the ear on the Hydraulic Base and Hydraulic Press using mate or insert constraints.

6. Apply a –90 degree angle constraint to the inside faces of the Hydraulic Base and Hydraulic Press as shown in Figure 5.93.

 Note: The angle may need to be 90 degrees based on how you placed and constrained the components.

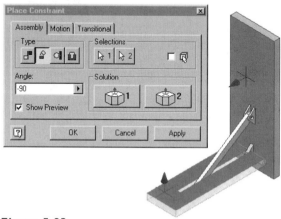

Figure 5.93

7. Drive the angle constraint between –90 and –180 degrees. Figure 5.94 shows the filled in Drive Constraint dialog box. Create an animation file and when done, compare your file against the file \Inventor Book\Hydraulic Cylinder Adapt.avi that is included on the CD.

 Note: The angle values may be positive.

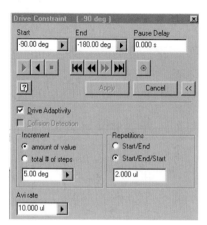

Figure 5.94

8. Save the assembly as \Inventor Book\Hydraulic Cylinder.iam

9. Create a presentation file with an exploded view that shows how the parts assemble (use your own values) and then create an animation file. For comparison, you can look at the file \Inventor Book\Hydraulic Cylinder Assembly.avi that is included on the CD.

REVIEW QUESTIONS

1. The only way an assembly can be created is by placing existing parts into it.   True or False?
2. Explain what the top down and bottom up assembly techniques refer to.
3. An occurrence is a copy of an existing component.   True or False?
4. Only one component can be grounded in an assembly.   True or False?
5. Inventor does not require components in an assembly to be fully constrained.   True or False?
6. Hole features cannot be adaptive.   True or False?
7. A sketch must be fully constrained to adapt.   True or False?
8. What is the purpose of creating a presentation file?
9. A presentation file is associated to the assembly file that it is based on.   True or False?
10. When creating drawing views from an assembly, you can create views from multiple presentation views or design views.   True or False?

CHAPTER 6

Advanced Dimensioning, Constraining, and Sketching Techniques

To this point in the book you have learned how to create sketches using step-by-step operations that have given you a good foundation of knowledge on how Inventor operates. In this chapter you will learn how to create more complex sketches and you will learn about tools that will reduce the number of steps to create a constrained sketch, use construction geometry to better control the sketch, set up relationships between dimensions, and create parts that are driven from a table.

CHAPTER OBJECTIVES

After completing this chapter, you will be able to:

- use construction geometry to help constrain sketches.
- create and constrain an ellipse.
- understand the type of splines that can be created.
- pattern sketches.
- copy sketches.
- share a sketch.
- utilize both the symmetry constraint and mirror tool.
- slice the graphics screen.
- project edges onto the current sketch.
- sketch on another parts face.
- use automatic dimensioning.
- create relationships between dimensions.
- create parameters.
- create a part that is driven from an Excel spreadsheet.

- create iParts.
- create a 2-D layout of an assembly.

CONSTRUCTION GEOMETRY

Construction geometry can help you create sketches that would be difficult to create without their aid. By default when you are sketching you are creating sketches that have a normal style, meaning that the sketch geometry will be used to create the component. There are four different styles that are available in Inventor: normal, construction, reference, and centerline. Reference geometry is created when you project edges from another part. Projecting edges will be covered later in this chapter. Geometry that uses the construction style can be constrained and dimensioned like normal geometry, but when the sketch is turned into a feature, the construction geometry will not be seen in the part. Construction geometry can reduce the number of constraints and dimensions that are required to fully constrain a sketch and can be used to help define the sketch. For example, a construction circle inside a hexagon could drive the size of the hexagon. Without construction geometry, the hexagon would require six constraints and dimensions; it would require only three constraints and dimensions with construction geometry—the circle will have tangent or coincident constraints applied to it.

Construction geometry can be created by changing the style either before or after it is created. Before sketching, select Construction from the Style area on the Command Bar; or after creating the sketch, select the geometry that you want to be construction, and then select Construction from the Style drop list as shown in Figure 6.1. After turning the sketch into a feature, the construction geometry will disappear. When you edit a feature's sketch that was created with construction geometry, the construction geometry will reappear during editing and disappear again when the part is updated. Construction geometry can also be added or deleted to a sketch just like any geometry that has a normal style. In the graphics area, construction geometry will be lighter in color and thinner in width than normal geometry.

Figure 6.1

ELLIPSES

Ellipses can be created by selecting the Ellipse tool from the Circle fly out on the Sketch Panel Bar as shown in Figure 6.2. To create an ellipse, pick a center point for the ellipse and then pick a point on the first axis and another point that will lie on the ellipse. The ellipse can be trimmed, extended, and dimensioned. To dimension an ellipse, start the General Dimension tool from the Sketch Panel Bar and then pick the ellipse and position the mouse over the major or minor distances. Then repeat the process to dimension the other major or minor distance as shown in Figure 6.3.

Figure 6.2

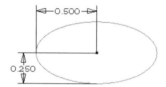

Figure 6.3

PATTERN SKETCHES

While creating a sketch that will have a rectangular or a polar pattern, you can choose to pattern the geometry in the sketch instead of after it is a feature. To pattern a sketch follow these guidelines.

- Pick either the Rectangular or Circular Pattern tool from the Sketch Panel bar as shown in Figure 6.4.
- Then the Rectangular or Circular Pattern dialog box will appear. Figure 6.5 shows the Rectangular Pattern dialog box.
- Pick the geometry button and select the geometry that will be patterned.
- Define the axis or direction(s) of the pattern.
- Type in the information for the count, spacing, and angle.
- At the bottom of the dialog box there is an option to suppress. Pick the suppress button and then select an occurrence(s) in the graphics window. To un-suppress an occurrence, pick the suppress button and then select an occurrence(s) in the graphics window. The occurrence will be displayed in a hidden style.
- Another option is to make the pattern associative or non-associative. An associative pattern will reflect any changes made to the geometry that was patterned. If non-associative is selected, any changes made to the geometry that is patterned will not be reflected in the patterned objects.
- The fitted option specifies whether the pattern occurrences will be equally fitted within the specified angle or distance. If fitted is not checked, the pattern spacing measures the angle or distance between occurrences.
- Pick the OK button to create the sketch pattern.

Once the pattern sketch has been created, the pattern will be grouped together. However, individual patterned occurrences can be deleted. The pattern sketch can be edited by right clicking on an occurrence of the pattern in the Graphics window and select Edit Pattern from the context menu. The same pattern dialog box will appear that was used to create it. Make changes as needed and then pick the OK button.

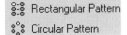

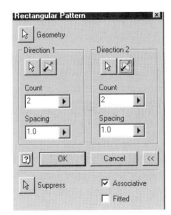

Figure 6.4

Figure 6.5

SHARED SKETCHES

When creating a part that would use the same sketch geometry, same dimensions, and would lie on the same face of the part, you can share the sketch instead of recreating it. A shared sketch is a copy of the original sketch and is placed above the original feature in the browser. The shared sketch has the same name or number of the original sketch. The visibility of the shared sketch will be turned on. Any modifications to a shared sketch will be updated to all the features that use the shared sketch. Once a sketch has been shared, the copy can be selected as the profile for additional sketched features. There is no limit to the number of features that can use the same shared sketch. The visibility of the shared sketch can be turned on and off as needed by right clicking on the shared sketch in the Browser and picking Visibility. To share a sketch, follow these steps.

1. In the Browser locate the feature that contains the sketch you want to share.
2. To the left of that feature, pick the plus sign to expand to the sketch.
3. Right click on the sketch icon and select Share Sketch as shown in Figure 6.6.

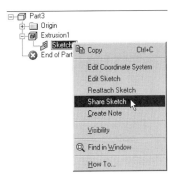

Figure 6.6

4. Issue the feature tool that you need (extrude, revolve, etc.), and then select the sketch as the profile for a sketched feature.

 Note: A shared sketch that is consumed by another feature cannot be deleted. By default a shared sketch's visibility is turned on. It can be turned on and off as needed by right clicking on its name in the Browser and selecting Visibility from the context menu.

MIRROR SKETCHES AND SYMMETRY CONSTRAINT

If you are creating symmetrical parts, you can utilize the Mirror sketch tool and the Symmetry constraint to reduce the number of dimensions and constraints that will be required to fully constrain the sketch. The Mirror tool will automatically apply a symmetry constraint(s) between the selected geometry and the mirrored geometry. Follow these guidelines to create a mirrored sketch:

- First analyze what the finished sketch will look like and determine where the line of symmetry will be. Draw half of the finished sketch and the line of symmetry. The line of symmetry must be a single line but its style can be normal, construction, or centerline.

- Constraints and dimensions can be added before or after the mirror operation.

- Pick the Mirror tool from the Sketch Panel bar as shown in Figure 6.7.

- The Mirror dialog box will appear. By default the Select option is active; in the graphics area select the objects that will be mirrored.

- Pick the Mirror line option and in the graphics area select the line that will be the line of symmetry.

- To complete the operation, pick the Apply button.

- A symmetry constraint will automatically be applied between both halves. Any change in a dimension or constraint to the first half will be reflected in the mirrored side. The symmetry constraint can be deleted like any other sketch constraint.

A symmetry constraint can also be added manually between two sets of objects. The object(s) must be the same object (line, arc, circle, etc.) and must lie on opposite sides of the symmetry line that must exist. After the geometries are drawn, add the Symmetry constraint from the Sketch Panel Bar under the sketch constraints fly out as shown in Figure 6.8. Then select the two geometries that will be symmetrical and select the line of symmetry.

Figure 6.7

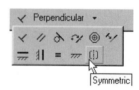

Figure 6.8

Advanced Dimensioning, Constraining, and Sketching Techniques 261

TUTORIAL 6.1—ELLIPSES, PATTERN SKETCHES, SHARED SKETCHES, AND THE MIRROR TOOL

1. Start a new part file based on the default Standard.ipt file.
2. Draw a rectangle and an ellipse and dimension them as shown in Figure 6.9.

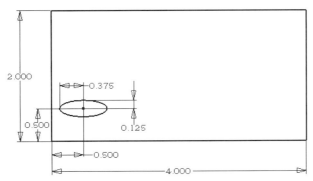

Figure 6.9

3. Use the Rectangular Pattern tool and pattern the ellipse as shown in Figure 6.10; pick the bottom horizontal line as Direction 1 and the vertical line as Direction 2 as shown with the arrows.

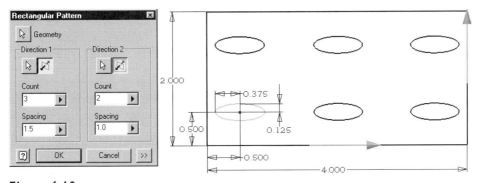

Figure 6.10

4. Change the minor dimension of the original ellipse from .125 to .06.
5. Edit the Rectangular Pattern by right clicking on one of the patterned occurrences and pick Edit Pattern from the context menu. Change the Count of Direction 1 to 2 and its spacing to 3 inches and then pick the OK button to complete the edit. When done your sketch should resemble Figure 6.11.

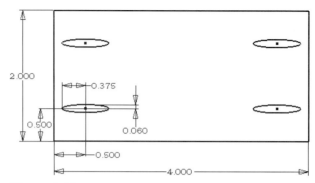

Figure 6.11

6. Close the file without saving the changes.
7. Start a new part file based on the default Standard.ipt file.
8. Draw and dimension a rectangle that is 3 inches x 1 inch.
9. Change to the default isometric view and extrude the rectangle 3 inches in the default direction.
10. Share the sketch of the rectangle by expanding Extrusion1 and right click on the sketch icon and select Share Sketch from the context menu.
11. Extrude the shared sketch with a taper of 15 and a distance of 3 inches. Reverse the extrusion direction so it is going into the screen.
12. Change the horizontal dimension of the shared sketch from 3 inches to 6 inches and update the part. When done your screen should resemble Figure 6.12.

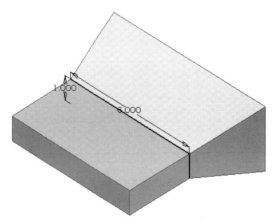

Figure 6.12

13. To turn off the visibility of the share sketch, right click on the name Sketch1 in the Browser and pick Visibility from the context menu to uncheck it.
14. Close the file without saving the changes.
15. Start a new part file based on the default Standard.ipt file.
16. Draw the sketch as shown in Figure 6.13. Make the left vertical line a centerline line by selecting it, and from the Style drop list on the Command Bar pick Centerline.
17. Pick the Mirror tool and then select all the geometry to mirror except the centerline.
18. Pick the Mirror Line button in the dialog box and select the centerline.
19. Pick the Apply button to create the mirror.
20. Pick the Done button to close the dialog box.
21. Select a point and drag it. The other side will move because of the symmetry constraint.
22. Add some dimensions and again the other side will update to reflect the change.

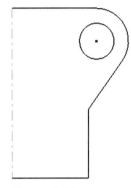

Figure 6.13

23. Close the file without saving the changes.
24. Start a new part file based on the default Standard.ipt file.
25. Draw the sketch as shown in Figure 6.14. Make the middle vertical line a centerline line by selecting it and pick Centerline from the Style drop list on the Command Bar.
26. Apply symmetry constraint(s) between the left and right side of your sketch. Use the centerline as the symmetry line.
27. Add some dimensions and the other side will update to reflect the change.
28. Close the file without saving the changes.

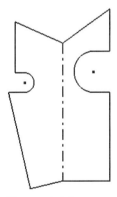

Figure 6.14

SLICE GRAPHICS

When creating parts, you may want to sketch on a plane that is difficult to see because other features are obscuring the view. The Slice Graphics option will temporarily slice away the portion of the model that obscures the plane on which you want to sketch. Follow these steps to temporarily slice the graphics screen. Figure 6.15 shows the Slice Graphics option active on a part.

1. Rotate the model so the plane of the active sketch faces you. The side facing you will be removed.

 Note: The plane must be the active sketch.

2. Right click and select Slice Graphics from the context menu as shown in Figure 6.15 or select Slice Graphics from the View pull down menu. The model will be sliced on the active sketch.

3. Use sketch tools from the Sketch toolbar to create geometry on the active sketch.

4. To restore the sliced graphics, right click and select Slice Graphics, select Slice Graphics from the View pull down menu, or click the Sketch or Return button from the Command Bar to end the sketch.

Figure 6.15

 Note: When working in the assembly environment notice that there are different slice graphics tools available from the Assembly toolbar.

PROJECT EDGES

When working on a component, you may want to transfer edges, vertices, work features, curves, or silhouettes from existing sketches onto the active sketch. The default center point can also be projected onto the current sketch to constrain it to the origin of the coordinate system. The Project tool will project selected edges, vertices, work features, curves, or silhouette edges of another part in the assembly, or of other features in the same part to the active sketch. If the setting Enable Associative Edge/Loop Geometry Projection During In-Place Modeling is checked from the Assembly tab in the Application Options dialog box as shown in Figure 6.16, the projected edges and vertices will be adaptive (dependent) on the model from which they are projected and maintain the relationship when the originating geometry changes. If this setting is not checked, the projected geometry will not be associative to the geometry that it was projected from. There are three project tools that are available from the Sketch Panel Bar as shown in Figure 6.17. The three tools are Project Geometry, Project Cut Edges, and Project Flat Pattern. The Project Geometry tool is used to project geometry from a sketch or feature onto the active sketch. The Project Cut Edges tool is used to project edges that lie on the section plane of a part on the active sketch and the geometry is only projected if the uncut part would intersect the sketch plane. The Project Flat Pattern tool projects a selected face or faces of the flat pattern onto the sheet metal part. Chapter 8 will cover creating sheet metal parts.

To project geometry, follow these guidelines.

- Click the Sketch tool, then pick a face or work plane to set create an active sketch.
- Pick the Project Geometry tool from the Sketch Panel Bar.
- Select the geometry to project onto the active sketch.
- To exit the operation press the ESC key or pick another tool.

To project cut edges, follow these guidelines.

- In an assembly with one or more components, create a section view, making sure that all components you need are visible.
- Create a new part whose sketch lies on the sectioned plane.
- Pick the Project Cut Edges tool from the Sketch toolbar.
- Select the geometry to project onto the active sketch.
- To exit the operation press the ESC key or pick another tool.

 Note: If the projected geometry is adaptive to another part, the adaptive link between a projected edge and its parent part in the assembly can be broken by expanding the adaptive part in the Browser. Right click on the sketch's name and uncheck Adaptive. The parts will then be independent of one another.

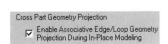

Figure 6.16

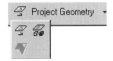

Figure 6.17

SKETCH ON ANOTHER PART'S FACE

In chapter 5 you learned that when a new part is created in an assembly you have an option to create the first sketch on a face or plane and have an assembly constraint automatically applied between the selected face on the other part and the first sketch. This relationship constrains the sketch to the selected face; if the face moves on the selected part so will the sketch. The option for creating a sketch on a face or plane on another part is not limited to the first sketch. Anytime a sketch is created, it can be placed on a face or plane of another part. This sketch will also be associated to the selected face or plane. When creating a sketch, pick the face or plane on the part that you want the sketch to be created on; then the sketch will be constrained to the selected face or plane of the other part. If the other part's face or plane moves, so will the sketch.

TUTORIAL 6.2—PROJECTING EDGES AND SKETCHING ON ANOTHER PART'S FACE

In this tutorial you will create a part by projecting geometry from a part to create another part. Then you will work on another assembly and create a feature on an existing part by creating a sketch on another part. Before working on this tutorial, ensure that the following settings from the Application Options dialog box from the Assembly tab are checked

- Mate Plane
- Adapt Feature
- Enable Associative Edge/Loop Geometry Projection During In-Place Modeling

1. Start a new Assembly file based on the default Standard.iam file.
2. Pick the Create Component tool from the Assembly Panel Bar. From the Create In-Place Component dialog box type in Base Plate for the New File Name and then select OK.
3. Sketch, constrain, and dimension the geometry as shown in Figure 6.18.

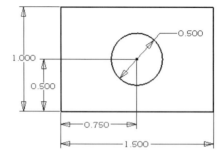

Figure 6.18

4. Change to an isometric view by selecting the right mouse button while in the graphics area and then selecting Isometric View from the context menu.
5. Extrude the sketch .1 inch in the default direction. For the Profile select the rectangle, but not in the middle of the sketch. The circle will become a void in the part.
6. Make the top-level assembly active by either picking the Return tool from the Command Bar or double click on the top-level assembly's name in the Browser.
7. Pick the Create Component tool from the Assembly Panel Bar. From the Create In-Place Component dialog box type in Projected Plate for the New File Name and check the Constrain sketch plane to selected face box. Pick OK in the dialog box and then select the left front face of the Base Part. A flush constraint will be applied and a sketch will become active on the selected plane.
8. Pick the Project tool from the Sketch Panel Bar and then pick a point on the middle of the face as shown in Figure 6.19.

Figure 6.19

9. Do *not* add any constraints or dimensions to the sketch.
10. Extrude the sketch .2 inches in the default direction. Again select the rectangular portion of the sketch for the profile. Do not include the circle.
11. Make the top-level assembly active by either picking the Return tool from the Command Bar or double click on the top-level assembly's name in the Browser.
12. Make the Base Plate active.
13. Edit the sketch and change the dimension of the circle to .25 and then pick the Update tool from the Command Bar. Both the diameter in the sketch and Projected Part will be updated to reflect the new value as shown in Figure 6.20.

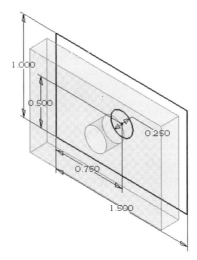

Figure 6.20

14. Make the top-level assembly active by either picking the Return tool from the Command Bar or double click on the top-level assembly's name in the Browser.
15. Close the file without saving the changes.
16. Open the assembly \Inventor Book\Part Sketch Assembly.iam
17. Make the part Part Sketch Base (orange part) the active part.
18. Pick the Sketch tool from the Command bar and pick the front of the angle face of the green part as shown in Figure 6.21.

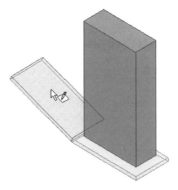

Figure 6.21

19. Pick the Project tool from the Sketch Panel Bar and then pick the left and bottom edge of the top of the angle as shown in Figure 6.22.

Figure 6.22

20. Sketch, constrain, and dimension the circle as shown in Figure 6.23. The horizontal and vertical dimensions are to the projected lines.

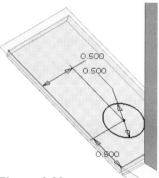

Figure 6.23

21. Extrude the circle using the To Extents and select the back face of the Part Sketch Base as shown in Figure 6.24.

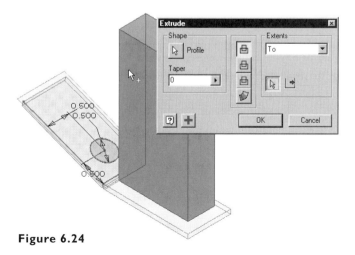

Figure 6.24

22. Make the top-level assembly active by either picking the Return tool from the Command Bar or double click on the top-level assembly's name in the Browser.
23. Change the value of the assembly Angle constraint between the parts Part Sketch.ipt:1 and Part Sketch.ipt:2 to −45 degrees.
24. To better see the new angle, change to the isometric viewpoint so the assembly resembles Figure 6.25.
25. Continue to try new values for the Angle constraint and watch the features angle change.
26. When done close the file without saving the changes.

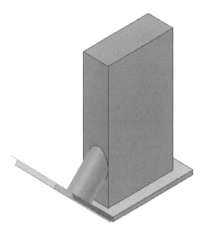

Figure 6.25

AUTO DIMENSION

Adding constraints and dimensions or removing dimensions to a sketch can be a time consuming task. To automate this process you can use the Auto Dimension tool, to automatically create or remove dimensions or add constraints to selected geometry. Before using the Auto Dimension tool, critical constraints and dimensions should be applied. The Auto Dimension tool will not override or replace any existing constraint or dimension. Select the Auto Dimension tool from the Sketch Panel Bar as shown in Figure 6.26. Then the Auto Dimension dialog box will appear as shown in Figure 6.27. In the lower left corner of the dialog box, the number of constraints and dimensions required to fully constrain the sketch will appear. Determine if you want to create dimensions or constraints or remove the dimensions/constraints that were previously added using the Auto Dimension tool. Then check the Dimensions and/or Constraints box. Select the Curve option and then in the graphics area select individual objects or window or cross the geometry that you want to work with. Pick the Apply button to create the dimensions and/or constraints to the selected curves. Or pick the Remove button to delete the selected dimensions. After the dimensions are placed you can change their values by double clicking on the numbers and entering in a new value.

 Note: If you use the Auto Dimension tool on the first sketch in the part, two dimensions or constraints will be required to fully constrain the sketch. Use the Fix constraint to remove these two dimensions or constraints.

Figure 6.26

Figure 6.27

DIMENSION DISPLAY, RELATIONSHIPS, AND EQUATIONS
DIMENSION DISPLAY

In chapter 1 you learned how to create independent dimensions that had no relationship to other dimensions. When creating parts, you may want to set up relationships between dimensions. For example, the length of a part may need to be twice that of its width, or a hole may always need to be in middle of the part. In Inventor there are a few different methods that can be used to set up relationships between dimensions. The first method uses a label that is given automatically to each dimension. Each label starts with the letter "d" and is given a number, for example "d0" or "d27." The first dimension created is given the label "d0" and each dimension that follows sequences up one number at a time. If a dimension is erased, the next dimension does *not* go back and reuse the erased value. Instead, it keeps sequencing from the last value on the last dimension created. When working, you may want to see the dimension's underlying label or label and value. There are three display types that can be set: Show Value, Show Name, and Show Expression. After you select a dimension display style, all the dimensions on the screen will change to that style. As dimensions are created, they will reflect the current dimension display style. To change the dimension display style, right click on a dimension in the graphics window and then pick a style from the context menu as shown in Figure 6.28. The two other display options will be available in the menu.

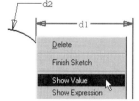

Figure 6.28

Show Value: The default dimension display style. When selected, the actual value of the dimensions will be displayed on the screen.

Show Name: Displays the dimensions as the dimension number or actual parameter name, for example d12 or Length.

Show Expression: The dimensions on the screen will change to d# = value, showing each actual value, for example, d7 = 4.50 or Length = 5.5 inches.

DIMENSION RELATIONSHIPS

To set a relationship between the dimension you are creating and an existing dimension, in the Edit Dimension dialog box you can type in the d number of the other dimension, or in the graphics area pick the dimension that you want to set the relationship to. Figure 6.29 shows the Edit Dimension dialog box after selecting the .750 dimension that the relationship will be related to.

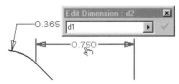

Figure 6.29

EQUATIONS

Equations can also be used whenever a value is required. A couple of examples would be (d9/4)*2 or 65 inches + 19 mm. When creating equations Inventor allows Prefixes, Precedence, Operators, Functions, Syntax, and Units. To see a complete listing of valid options, use the online help and navigate to the topic Equations and the title Edit box reference. Numbers can be entered with or without units; when no unit is entered the default unit will be assumed. As you enter an equation Inventor is calculating it. An invalid expression will be shown in red and a correct expression will be shown in black.

To create an equation in an edit box follow these guidelines:

- Click in the edit field.
- Type any valid combination of numbers, parameters, operators, or built-in functions.
- Press Enter or click OK to accept the expression.

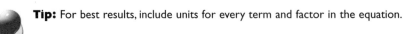

Tip: For best results, include units for every term and factor in the equation.

PARAMETERS

Another method of setting up relationships between dimensions is to use parameters. A parameter is a user-defined name that is assigned a numeric value either explicitly or through equations. Multiple parameters can be used in an equation and parameters can be used to define another parameter such as depth = length – width.

A parameter can be used anywhere a value is required. A parameter can be defined by any name except d# where # is any number (the d-number combinations are used for dimensions). There are three types of parameters: Model parameters, User parameters, and Linked parameters that are defined from a spreadsheet.

- Model parameters are automatically created and assigned a name when a sketch dimension, feature, or assembly constraint is created.
- A User parameter is manually created by the user in the Parameter dialog box.
- A Linked parameter, via a spreadsheet, is created in Excel.

User parameters are only used in the file that they are created in. If a spreadsheet is used to link parameters, there is no limit to the number of part files that can reference the same spreadsheet. If the same User parameter name is used over and over, it should be defined in a template file. When new parts are created that are based on that template, the parameters will be defined.

When creating parameters follow these guidelines.

- Assign meaningful names to parameters; other people may edit the part file and will need to determine your thought process.
- The parameter name cannot include spaces, mathematical symbols, or special characters. These can be used to define the equation.
- The parameter name must include at least one alpha character besides the letter d. For example W1 or Width1 would be valid. 123 would be an invalid parameter name.
- When defining a parameters equation you cannot use the parameter name to define itself. For example Length = Length/2 would not be valid.

To create a User parameter, pick the Parameters tool from the Standard toolbar as shown in Figure 6.30. Then the Parameters dialog box will appear as shown in Figure 6.31. The dialog box is divided into two sections: Model Parameters and User Parameters. The Model Parameters section is automatically filled with the values from dimensions or assembly constraints that were used in the file. The User parameters will manually be defined. For both types of parameters, the names and equation can be changed and comments can be added. To edit a name or equation

Figure 6.30

or add a comment, double click in the cell and type the new information. User parameters can be created by picking the Add button from the bottom of the dialog box and then type in the information in each of the cells. The column names for both Model and User Parameters are the same and are defined in the following section.

Parameter Name: In this cell the name of the parameter will be displayed. To change the name of an existing parameter, click in the box and enter a new name or when creating a new User Parameter after picking the Add button, type in a new parameter name. When you update the model, all dependent parameters update to reflect the new name.

Units: In this cell enter a new unit of measurement for the parameter.

Equation: In this cell the equation will be displayed; from the equation, the value of the parameter will be determined. If the parameter is a discrete value, the value is displayed, rounded to match the precision setting for the model. To change the equation, click the existing equation and enter the new equation.

Value: In this cell, the calculated value of the equation will be displayed in full precision.

Export Parameters Column: Check this box if you want the parameter to be exported to the Custom tab of the component properties dialog box and the parameter will also be available in the bill of materials and parts list Column Chooser dialog boxes.

Comment: If desired, type a comment for the parameter in this cell. To add a comment, click in the cell and enter the comment.

After creating the parameter(s), you can type its name anywhere a value is required; or when editing a dimension, pick the arrow, and from the context menu pick List Parameters as shown in Figure 6.32. All the available parameters will be displayed in a list; from that list pick the desired parameter. When typing in the parameter names, they are case sensitive. An invalid parameter name will be displayed in the color red and valid names will be displayed in the color black. After editing a parameters value, changes will occur after you pick the Update button from the Command Bar.

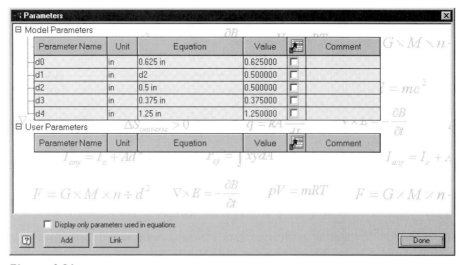

Figure 6.31

Figure 6.32

PARAMETERS LINKED TO A SPREADSHEET

If you want to use the same parameter name and value in many different parts, you can create a spreadsheet using Microsoft Excel. Then you can either embed or link the spreadsheet into assembly or part files. When an Excel spreadsheet is embedded, there is no link between the Excel spreadsheet and the Parameters in Inventor. When an Excel spreadsheet is linked to the part file, any changes in the Excel spreadsheet will also update the parameters in the linked Inventor file. Parameters driven from a spreadsheet are used in the modeling process just like any other parameter can be used.

When creating a Microsoft Excel spreadsheet with parameters, follow these basic rules.

- The data can start in any cell of the spreadsheet but must be specified when linking or embedding the spreadsheet.
- The data items can be in rows or columns, but they must be in the correct order: parameter name, value or equations, unit of measurement and, if needed, a comment.
- The parameter name and value are required; the other items are optional.
- The parameter name cannot include spaces, mathematical symbols, or special characters. These can be used to define the equation.
- If you do not specify the units of measurement for a parameter, the default units for the model are assigned when you use the parameter. To create a parameter without units, enter UL (unitless) in the units cell.
- You can include column or row headings or other information in the spreadsheet, but they must be outside the block of cells that contains the parameter definitions.

Figure 6.33 shows an example of three parameters that were created in rows with a name, equation, unit, and comment.

	A	B		D
1	Length	5	in	Length of Plate
2	Width	3	in	Width of Plate
3	Depth	0.25	in	Depth of Plate

Figure 6.33

After the Excel spreadsheet is created and saved, you can link to it from within Inventor by picking the Parameters tool from the Standard toolbar as you did to create a parameter. From the bottom of the Parameter dialog box pick the Link button and the Open dialog box will appear like what is shown in Figure 6.34. Navigate to and select the Excel spreadsheet file and then type in the starting cell and select if the spreadsheet will be Linked or Embedded. With Embedded, there is no link between the Excel spreadsheet and the Parameters that will be created in Inventor. When an Excel spreadsheet is linked to the part file, any changes in the Excel spreadsheet will also update the parameters in the linked Inventor file.

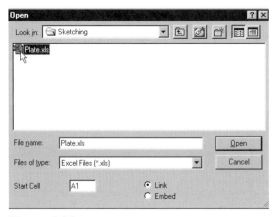

Figure 6.34

After selecting the file and options, pick the Open button and a table showing the parameters is added to the Parameters dialog box like what is shown in Figure 6.35. You cannot edit the parameters that are driven by a spreadsheet from the Parameters dialog box. If the spreadsheet is linked, you can open the spreadsheet in Microsoft Excel and edit it; or from within Inventor, open the part file that uses the spreadsheet, and from the Browser expand the 3rd Party folder and double click the spreadsheet to edit it. The spreadsheet will open in a Microsoft Excel window for editing. If the spreadsheet is embedded, from within Inventor open the part file that uses the spreadsheet, and from the Browser expand the 3rd Party folder and double click the spreadsheet to edit it. The spreadsheet will open in a Microsoft Excel window for editing. After editing the Excel spreadsheet, save the changes and make Inventor the current application; then pick The Update tool from the Command Bar.

 Note: When typing a parameter name where a value is requested, the upper and lower case of the letters must match the parameter name. Duplicate parameter names are not allowed. Model, User, and Spreadsheet driven parameters must have unique names.

Advanced Dimensioning, Constraining, and Sketching Techniques 277

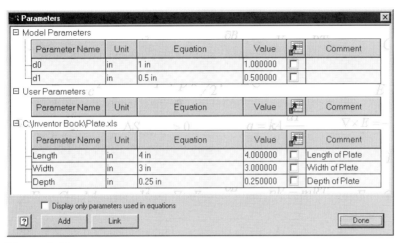

Figure 6.35

TUTORIAL 6.3—AUTO DIMENSION, RELATIONSHIPS, PARAMETERS, AND SPREADSHEETS

1. Start a new part file based on the default Standard.ipt file.

2. Draw the geometry that is shown in Figure 6.36. The size should be roughly 1 inch wide by 1.5 inches high.

3. Pick the Auto Dimension tool, then from the Auto Dimension dialog box pick the Curves button and select all the geometry. Then with both the Dimensions and Constraints options checked, pick the Apply button to create the dimensions. Then pick the Done button to exit the operation, and then your sketch should resemble Figure 6.37. Your dimensions may be in different places depending upon how you created your geometry.

4. Erase and recreate any dimensions so you have four dimensions that resemble Figure 6.37.

5. To fully constrain the sketch, add a fix constraint to the lower left corner.

6. Change all the dimensions values to match Figure 6.38.

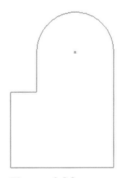

Figure 6.36

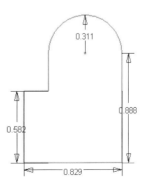

Figure 6.37

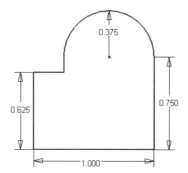

Figure 6.38

7. Now you will set up relationships between the dimensions. Edit the .375 radius dimension of the arc and for its value select the 1.000 horizontal dimension as shown in Figure 6.39. Then type in /4 after the d#. For example, d1/4. The 1 may be a different number depending upon how your geometry was created.

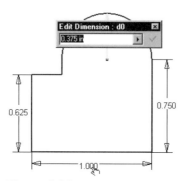

Figure 6.39

8. Now edit the .625 vertical dimension and for its value select the .750 vertical dimension. Then type in /2 after the d#.

9. Change the value of the 1.000 horizontal dimension to 2.000 and the .750 vertical dimension to .500. When done your sketch should resemble Figure 6.40.

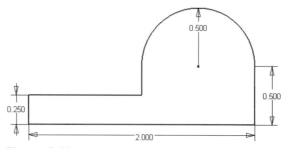

Figure 6.40

10. Now you will create a couple of parameters and drive the sketch from them. Pick the Parameters tool from the Standard toolbar.

11. In the Parameter dialog box, create a User parameter by picking the Add button and then type in the following information.

    Parameter Name = Length
    Units = in
    Equation = 3
    Comment = Bottom Length

12. Create another user parameter by selecting the Add button and type in the following information. When done the User Parameter area should resemble Figure 6.41.

    Parameter Name = Height
    Units = in
    Equation = Length/2
    Comment = Height is half of the length

13. Close the Parameter dialog box by clicking Done.

14. Change the dimension display to Show Expression by right clicking on a dimension and select Show Expression from the context menu.

Parameter Name	Unit	Equation	Value		Comment
Length	in	3 in	3.000000	☐	Bottom Length
Height	in	Length / 2 ul	1.500000	☐	Height is half of the Length

Figure 6.41

15. Edit the bottom horizontal dimension and change its value to the parameter "Length." Double click on the dimension and in the Edit Dimension dialog box, pick the arrow; from the context menu pick List Parameters and then pick Length from the list. Then edit the right vertical dimension and change its value to Height. When done your screen should resemble Figure 6.42.

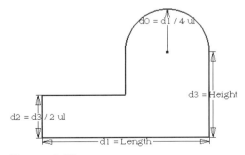

Figure 6.42

16. Pick the parameters tool from the Standard toolbar. From the User Parameters section double click on the equation cell of the parameter name Length and change its value to 1.5.

17. Repeat the same steps for the parameter name Height and change its value to 3 and the comment field to Height of the right side. When done making the changes, the User Parameter area should resemble Figure 6.43.

Parameter Name	Unit	Equation	Value		Comment
Length	in	1.5 in	1.500000	☐	Bottom Length
Height	in	3 in	3.000000	☐	Height of the right side

Figure 6.43

18. To complete the changes pick the Done button, and then the sketch should update to reflect the changes.

19. Now you will create a spreadsheet that has two parameters. Create an Excel spreadsheet that has the column names Parameter Name, Equation, Unit, and Comment along with the data shown in Figure 6.44.

Parameter Name = BaseExtrusion
Units = in
Equation = 1.25
Comment = Extrusion distance for the base feature

Parameter Name = Draft
Units = deg
Equation = 5
Comment = Draft for all features

	A	B	C	D
1	Parameter Name	Equation	Unit	Comment
2	BaseExtrusion	1.25	in	Extrusion distance for the base feature
3	Draft	5	deg	Draft for all features

Figure 6.44

20. Save the Excel spreadsheet as \Inventor Book\Chap6 Tutorial 3.xls.

21. Make Inventor the current application and then pick the Parameters tool from the Standard toolbar.

22. In the Parameter dialog box pick the Link button and then select the file "\Inventor Book\Chap6 Tutorial 3.xls."

Advanced Dimensioning, Constraining, and Sketching Techniques   281

23. For the start cell type in A2. If you fail to do this, no parameters will be found.

24. Then pick the Open button and a new spreadsheet area will appear in the Parameters dialog as shown in Figure 6.45. Then pick the Done button to complete the operation.

Parameter Name	Unit	Equation	Value		Comment
BaseExtrusion	in	1.25 in	1.250000		Extrusion distance for the base feature
Draft	deg	5 deg	5.000000		Draft for all features

Figure 6.45

25. Change to an isometric view.

26. Pick the Extrude tool from the Features Panel bar and for the Taper value type Draft or pick the arrow; from the context menu choose List Parameters and then the pick parameter Draft from the list and for Distance value use the parameter BaseExtrusion. To complete the operation, pick the OK button. Figure 6.46 shows the filled in Extrude dialog box and the finished part.

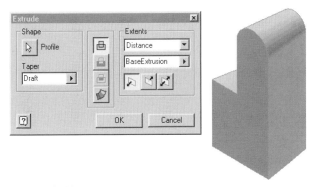

Figure 6.46

27. In the Browser expand the 3rd Party icon and either double click or right click on the name "Chap6 Tutorial 3.xls" and select Edit from the context menu as shown in Figure 6.47.

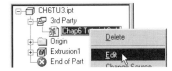

Figure 6.47

28. In the Excel spreadsheet make the following changes.

    For the Parameter Name "BaseExtrusion" change the Equation to 3
    For the Parameter Name "Draft" change the Equation to –3

29. Save the Excel spreadsheet and then Close Excel.

30. Make Inventor the current application and pick the Update tool from the Command bar. When done your part should resemble Figure 6.48.

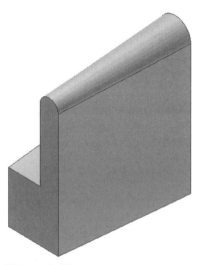

Figure 6.48

31. Save the file as \Inventor Book\Chap6 Tutorial 3.ipt
32. Close the file.

iPARTS

In the last few sections about parameters you learned you could only store one value for each parameter name. If you want to create a part that can have multiple values for the same dimension or parameter name, you can save these different configurations into an iPart. In industry these configured parts have been referred to as tabulated parts, charted parts, or a family of parts. iParts can also be used to create part libraries that will allow design data to be reused. iParts provides the ability for you to suppress specific features, add properties to the different members that are contained within the iPart, and allow user input for specific parameters within a predetermined range. After creating an iPart, it can be placed into an assembly where the specific configuration (or member) can be selected. There is no limit to the number of configurations of the same part that can be placed into the same assembly. The following sections will explain how to create and then place an iPart into an assembly.

An example of an iPart is a simple bolt. The bolt has a number of different sizes that are associated with it. An iPart allows you to define these different sizes and configurations (material, part properties, etc.) and have them reside within a single file. When placing the bolt in an assembly, you can select which member (or size) of the bolt (or iPart) that you want to use. The placed member can then be constrained similar to any other part file.

CREATING iPARTS

There are two types of iParts that can be created: Standard and Custom.

Standard iPart (Factories): Standard iPart values cannot be modified. When placing a standard iPart into an assembly, only the predetermined configurations can be selected. A standard iPart that is placed in an assembly *cannot* have features added to it.

Custom iPart (Factories): Custom iParts allow the user to place their own value for at least one variable. A custom iPart that is placed in an assembly *can* have features added to it.

To create an iPart follow these guidelines.

- Create an Inventor part or sheet metal part. Place dimensions on the geometry of the design that will be changed in the configuration. For easier creation of the configurations, use descriptive names for the values of the parameters. If parameter names are not used, you will need to determine what each parameter (or d#) represents within the parts geometry.
- Issue the iPart Author tool from the Standard toolbar as shown in Figure 6.49 to add the members or configurations of the iPart.

Figure 6.49

- The iPart Author dialog box will appear as shown in Figure 6.50. Follow these steps to create the iPart members.

    First, parameters and dimensions that will be configured need to be added to the right side of the dialog box (parameter list). All named parameters will automatically appear on the right side. To add a d# dimension to the list on the right side, select it from the left column and then pick the >> button as also shown in Figure 6.50. To remove a parameter from the right side of the dialog, select its name and pick the << button.

    After the parameters have been added, define the keys. Keys identify a column whose values are used to define the iPart configuration when the part is published (or placed) in an assembly. For example, setting a parameter as a primary key allows the designer placing the part to choose from all available values for that parameter in the selected items list. To specify the key order, click an item in the key column of the selected parameters list to define it as a key, or right click on the item

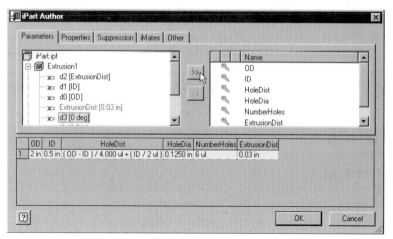

Figure 6.50

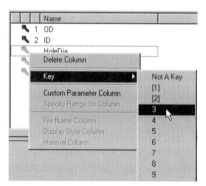

Figure 6.51

and select the key sequence number as shown in Figure 6.51. Selected keys are blue; items that are not selected as keys have dimmed key symbols. One primary key (Key [1]) is required, but you can add up to eight secondary keys. Custom columns cannot have a key assigned to them.

From the other tabs in the dialog box pick the specific tab to do the following. These items are not required to define the iPart factory.

> From the File Properties tab, select one or more file properties to include them in the iPart.

> From the Suppression tab, select a feature that should be suppressed. Features will be computed if the status is Compute and will be suppressed if the status is Suppress.

Advanced Dimensioning, Constraining, and Sketching Techniques 285

From the iMates tab, if needed, select one or more iMates to include in the iPart. iMates are included in the iPart if its status is Compute. If their status is suppressed, they are not included in the iPart.

From the Other tab, create custom values such as Color, Material, or Filename, if desired, to be included in the iPart.

Next add a configuration in the iPart table by right clicking in a row at the bottom of the dialog box and select Insert Row from the context menu as shown in Figure 6.52. Edit the cell contents as needed by picking in the cell and typing new values or information as needed. To delete a row, right click on the row number and pick Delete row from the context menu. Each row that is added in the bottom portion of the dialog box represents an additional member (or part) within the iPart factory.

Figure 6.52

- To allow the designer placing the iPart to specify the value of a given column, right click on the column name and select Custom parameter Column from the context menu as shown in Figure 6.53. Or to make a specific cell custom, right click in the individual cell and pick Custom Parameter Cell from the context menu. After designating a Custom Parameter Column or Cell, you can set a minimum and maximum range of values by right clicking in the column or cell and selecting Specify Range for Column or Range for Cell from the context menu. The Specify range dialog box will appear. Select the options as needed. Custom columns will be highlighted with a blue background.

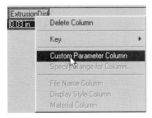

Figure 6.53

- Set the member that will be the default by right clicking on the row number and pick Set As Default Row from the context menu. The other options may be selected as needed.

- Save the iPart file to a location that will be available to all the users that need access to them. If the iPart is a standard part that will not be modified, it is better to save the iPart to a library. The library path name should be added to the project file(s) that will be used when the iPart is placed. The library directory where you save the iParts must have the same name as the factory library, preceded with an underscore character. When you add the location of an iPart factory as a library, you must add a second, special library for the iParts published from the factory. For example, if your factories are stored in a library named Bolts, you should define a library named _Bolts. Inventor will automatically store all iParts published by the factory in the library _Bolts. If you store an iPart factory in a workspace, a local search path, or a workgroup search path, the published iParts are stored in a subdirectory with the same name as the factory, located in the same directory as the factory.

 Note: It is recommended that you do not store iParts in the same folders as the iPart factories.

- To edit an iPart open the iPart file and in the Browser double click on the Table icon and the iPart Author dialog box will appear. Make changes as needed and then save the file. Changes to iParts that have been placed in assemblies will not automatically be updated. To update the iParts in an assembly, open the assembly and pick the Full Update tool from the Standard toolbar. iParts that need to be updated are marked with an update symbol in the Browser.

iPART PLACEMENT

To place a Standard iPart into an assembly follow these guidelines.

- Start a new assembly or open an existing assembly that the iPart will be placed in.

- Use the Place Component tool and in the Open dialog box navigate and select the iPart that will be placed in the assembly.

- Select the Open button and the Place Custom iPart or Place Standard iPart dialog box will appear. If there is a custom cell(s) or column(s) the Place Custom iPart dialog box will appear like what is shown in Figure 6.54.

- To select from the member list, pick the Table tab and then select the row that defines the part you want to place. Figure 6.54 also shows the Table tab with the arrow on the second row, when selected this will make that row active.

- If a custom part is being placed, pick the Keys tab and type in a value in the right side of the dialog box as shown in Figure 6.55. The value must fall within the limits set in the iPart factory, otherwise an alert will appear.

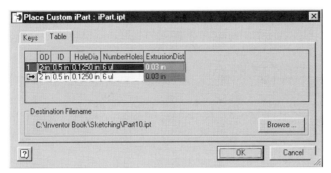

Figure 6.54

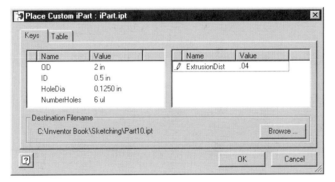

Figure 6.55

- In the graphics area place the part.
- Continue placing iParts and when done pick the Dismiss button.
- To change an iPart in an assembly to a different configuration, in the Browser expand the part and right click on the table name and choose Change Component. Then from the Table tab, select the new row.

Tip: Use the Parameters tool on the Standard toolbar to rename system parameters and to create unique parameter names before you use the iPart Author tool. Named parameters are automatically added to the table-driven list when you create the iPart Factory.

If no unit of measurement is specified in a cell, the default document units are used.

If you select the Material property from the Design Tracking Properties list, you must use the Material Column option on the Other tab to ensure that the current color is set to As Material.

In order for part properties to be used in drawings and bills of materials, you must include them in the iPart table, even if their values do not vary between configurations.

When changes are made to an iPart factory, the changes will not automatically update any iParts that are in assemblies. Open the assembly and pick the Full Update tool from the Standard toolbar.

A standard iPart that is placed in an assembly *cannot* have features added to it.

TUTORIAL 6.4—iPARTS

In this tutorial you will create an iPart with multiple configurations and then place them in an assembly. User parameters have been created in the part that you will turn into an iPart.

1. Open the file \Inventor Book\iPart Gasket.ipt
2. Start the iPart Author tool from the Standard toolbar.
3. There are two parameters that will not be changed by the user: HoleDist and ExtrusionDist. Remove them from the parameter list by selecting the name and then picking the << button. Repeat this for both HoleDist and ExtrusionDist. Figure 6.56 shows the HoleDist parameter being removed from the parameter list.

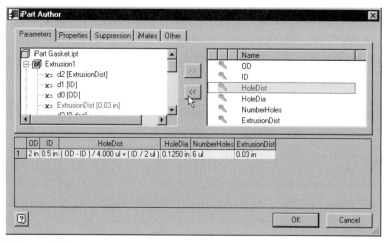

Figure 6.56

4. Insert two rows in the configuration area by right clicking on row number 1 and select Insert Row and then repeat this to insert another row.
5. In row 2 change the OD to 3 and the ID to .75.

6. In row 3 change the OD to 4, ID to 1 and HoleDia to .25. When done the iPart Author dialog box should resemble Figure 6.57.

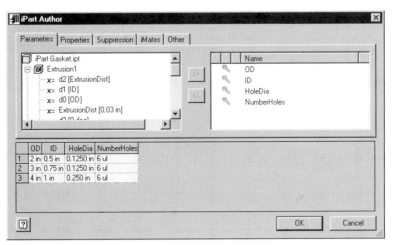

Figure 6.57

7. To allow the designer to input the number of holes that fall within a range, right click on the column name NumberHoles and select Custom Parameter Column from the context menu.

8. To specify the range right click in the column name NumberHoles and select Specify Range for Column from the context menu. Then in the Specify Range NumberHoles dialog box, change the minimum symbol to ≤ and the value to 3 ul. Then change the maximum symbol to ≤ and the value to 6 ul. When done the dialog box should resemble Figure 6.58. Click OK to close the Specify Range dialog box.

Figure 6.58

9. Select OK to complete the iPart Factory and then save the file.
10. Close the file.
11. Start a new Assembly file based on the default Standard.iam file.

12. Pick the Place Component tool from the Assembly Panel Bar and pick the file \Inventor Book\iPart Gasket.ipt. Pick the Open button and the Place Custom iPart dialog box will appear.

13. Pick the Table tab and select row number 2 as shown in Figure 6.59.

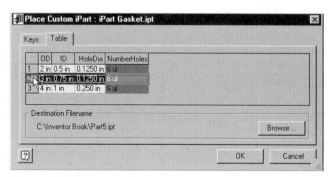

Figure 6.59

14. Pick the Keys tab and type in a value of 7 for the custom parameter NumberHoles on the right side of the dialog box and then press ENTER. An alert box will appear stating the valid range. Notice that the parameter is *red* because it is outside of the range.

15. Type in a value of 4 for the custom parameter NumberHoles and press ENTER and then the dialog box should resemble Figure 6.60. Pick a location in the graphics window to place the first part in the assembly.

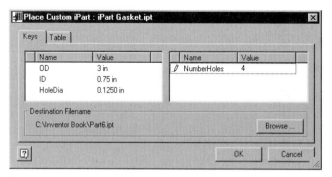

Figure 6.60

16. Pick the OK button to complete the operation.

17. Again pick the Place Component tool from the Assembly Panel Bar and pick the file \Inventor Book\iPart Gasket.ipt. Pick the Open button and the Place Custom iPart dialog box will appear.

18. Pick the Table tab and select row number 3.
19. Pick a point to the right of the first part. Then pick the OK button. When done your screen should resemble Figure 6.61.

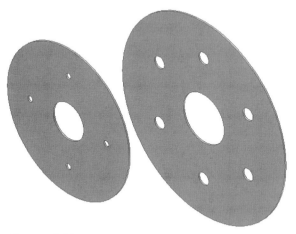

Figure 6.61

20. In the Browser expand the second part and right click on Table. From the context menu pick Change Component.
21. The Place Custom iPart dialog box will appear. Pick the Table tab and select row number 1.
22. Pick the Key tab and type in a value of 3 for the custom parameter NumberHoles and then select OK to complete the change. When done your screen should resemble Figure 6.62.
23. Close the assembly file without saving the changes.

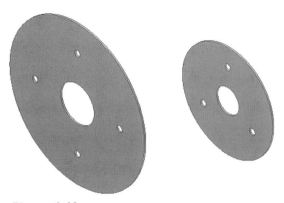

Figure 6.62

2-D DESIGN LAYOUT

In chapter 5 (Assemblies) you learned how to create 3-D parts and assemble them. If you are designing a mechanism and you want to test the function before giving the part form or detail (function before form), you can create 2-D parts (2-D design layout) and assemble them and then test the mechanism. If the mechanism works the way you expect, you can turn the 2-D parts into 3-D parts. The 2-D sketches do not need to be turned into 3-D parts. If the 2-D sketches represent purchased parts, you may choose to leave them in a 2-D state. When creating a 2-D layout, you will follow the same rules that you used to create 3-D parts. When working with design layouts, you can mix 2-D and 3-D parts and create individual part files that can be brought into an assembly. You can also create the parts while in an assembly or use a combination of both methods. Assembly constraints can be added to 2-D parts just like a 3-D part.

TUTORIAL 6.5—2-D DESIGN LAYOUT

In this tutorial you will create a simple 2-D lever and apply assembly constraints between the 2-D lever and an existing part. Assembly constraints have been applied between the two existing parts but the plate can still move up and down.

1. Open the assembly \Inventor Book\Clamp Assembly.iam.
2. Verify that the middle plate can move up and down by dragging it.
3. Create a new component in place. Give the part the name "2-D Lever" and if it is not already checked, check the Constrain sketch plane to selected face box.
4. Then pick the OK button and select the Blue work plane that is in the middle of the Clamp Bottom Plate.
5. Turn off the visibility of the work plane.
6. Draw two circles and dimension them as shown in Figure 6.63. Apply a vertical constraint between the centers of the two circles.

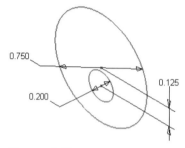

Figure 6.63

7. Draw a horizontal line from the center of the larger circle and apply a 1.5 dimension as shown in Figure 6.64.

Advanced Dimensioning, Constraining, and Sketching Techniques 293

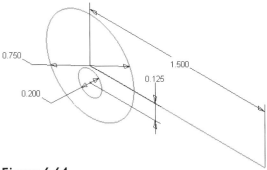

Figure 6.64

8. Make the top level of the assembly active.
9. Apply a mate constraint between the centerline of the hole in the back side plate of the Clamp Bottom Plate and the small circle on the 2-D Lever as shown in Figure 6.65.

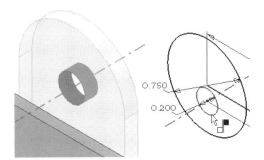

Figure 6.65

10. Apply an outside tangent constraint between the top of the Clamp Plate and the bottom of the large circle of the 2-D Lever as shown in Figure 6.66.

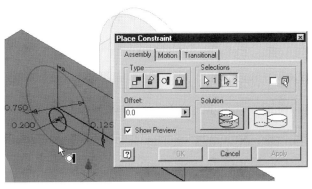

Figure 6.66

11. Test the mechanism by selecting the 2-D handle on the lever and rotate it. The plate should move up and down.

12. Save the assembly and the 2-D Lever.

13. Close the file.

PRACTICE EXERCISES

Follow the directions for the two exercises below. When you have completed the part, save the file(s) as noted.

Exercise 6.1—iPart-Rectangular Steel Tubing

In this exercise create an iPart for rectangular steel tubing. Use the following parameter names and default values.

WIDTH = 1.5 in
HEIGHT = .75 in
WALL = .083 in
LENGTH = 10 in

When sketching the rectangle for the tube, use the Two point rectangle tool and then use the offset tool to create the inside rectangle. The offset tool by default will offset all four lines of the rectangle, and when the offset value changes the offset relationship will be maintained to all four lines. When placing the dimensions, use the parameter *width* for the horizontal dimension, parameter *height* for the vertical dimension, and use the parameter *wall* for the offset dimension. Use the parameter *length* for the extrusion distance. When creating the iPart, use the values and place in the Key numbers as shown in Figure 6.67 or use the following chart for the values. Make the column *length* a custom column. After creating the iPart, try placing different configurations into an assembly.

	Width	Height	Wall	Length
1	1.5 in	0.75 in	.083 in	10 in
2	1.5 in	1.00 in	.065 in	10 in
3	2 in	1.00 in	.065 in	10 in
4	2 in	1.5 in	.083 in	10 in
5	2.5 in	1 in	.083 in	10 in

Save the finished iPart as \Inventor Book\EX6-1.ipt

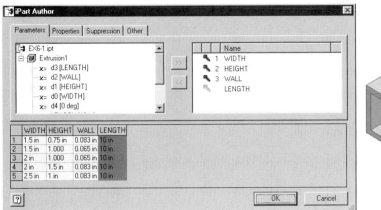

Figure 6.67

Exercise 6.2—2-D Layout

For this exercise the Base Cylinder has been created and a work plane that goes through the center of the cylinder has been created but its visibility is turned off. You will create two 2-D parts and then assemble both of them together.

1. Open the assembly \Inventor Book\EX6-2.iam.

2. Create a new part in the assembly named EX6-2 Lever. If it is not already checked, check the Constrain sketch plane to selected face box. Then when selecting the face for the sketch, in the Browser expand the part "EX6-2 Cylinder.ipt" and pick Work Plane1. Sketch and dimension the slot and circle as shown on the left side of Figure 6.68.

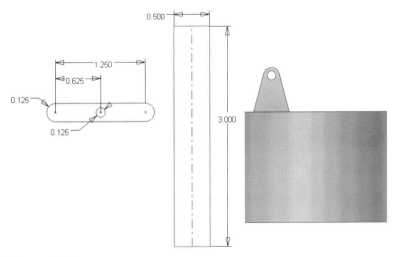

Figure 6.68

3. Create a new part in the assembly named EX6-2 Rod. If it is not already checked, check the "Constrain sketch plane to selected face" box. Then when selecting the face for the sketch, in the Browser expand the part "EX6-2 Cylinder.ipt" and pick Work Plane1. Sketch and dimension the rectangle and centerline as shown in the middle of Figure 6.68.

4. Change to an isometric view and assemble the parts together. Mate the centerlines between the circle in the middle of the Lever and the holes of the ears of the Cylinder, and between the centerline of the Rod and center of the Cylinder. Add a Tangent constraint between the right side arc of the Lever and the top horizontal line of the Rod. Figure 6.69 shows the parts assembled in a plan view.

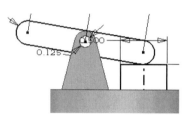

Figure 6.69

5. Drag the lever and the rod should go up and down. Drag the rod and the lever should go up and down. To verify your results, play the avi file \Inventor Book\EX6-2.avi.

6. Save the assembly and component parts.

7. Close the assembly file.

REVIEW QUESTIONS

1. Geometry that uses the construction style cannot be dimensioned to. True or False?

2. Modifications to a shared sketch will update all the features that use that shared sketch. True or False?

3. Slice Graphics will permanently slice away a portion of the model. True or False?

4. The Project tool can project vertices, work features, curves, or silhouette edges of another part in the assembly to the active sketch. True or False?

5. If the Auto Dimension tool is used on the first sketch in the part, the sketch will be fully constrained. True or False?

6. What is the difference between a Model Parameter and a User Parameter?

7. When creating parameters in a spreadsheet, the data items must be in the following order: parameter name, value or equations, unit of measurement and, if needed, a comment. True or False?

8. What is the difference between a Standard iPart and a Custom iPart?

9. Multiple versions of an iPart can be placed in an assembly. True or False?

10. When creating a design layout you *cannot* mix 2-D and 3-D parts. True or False?

CHAPTER 7

Advanced Modeling Techniques

Up to now, you have learned basic modeling techniques. In this chapter you will learn how to use advanced modeling techniques. Using advanced modeling techniques, you can create transitions between parts that would otherwise be difficult to create. Advanced features can be edited like other sketched and placed features, but typically require more than one unconsumed sketch in order to be created. In this chapter you will also be introduced to techniques that will help you be more productive in your modeling by learning how to copy features, reuse features that have been created using iFeatures, and create one-off designs using derived parts and additional functionality with derived assemblies. Surfaces are also available and can be used in conjunction with sketched features as a termination plane for an extrusion or to split a part.

CHAPTER OBJECTIVES

After completing this chapter, you will be able to:

- create Sweep features.
- create Coils.
- create Loft features.
- split a part or split faces of a part.
- copy features within a model.
- reorder model features.
- suppress features of a model.
- use Inventor Surfaces.
- create Ribs, Webs, and Rib Networks.
- create iFeatures.
- use Derived Parts and Assemblies.
- work with File Properties

SWEEP

In this section you will learn how to create sweep features. A sweep feature allows you to create a feature by *sweeping* a profile along a path. To create a sweep feature you need to have two unconsumed sketches. One sketch that represents the profile and a second sketch that represents the path that you would like the profile to be swept along. Using the sweep feature you can create a feature that follows either an irregular path or that runs along or close to the edge of a part. The path can be either open or closed. Handles, cabling, and piping are typical swept features.

As mentioned, sweep features are made up of two unconsumed sketches, the profile and the path (open or closed). After two unconsumed sketches have been created, select the Sweep tool from either the Features Panel Bar as shown in Figure 7.1, or the Features toolbar, and the Sweep dialog will open as shown in Figure 7.2. If two unconsumed sketches do not exist, Inventor will notify you that two unconsumed sketches are required.

Figure 7.1

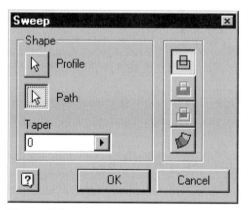

Figure 7.2

Following is a description of the sections in the Sweep dialog box.

SHAPE

Profile: Select this button to choose the sketch to sweep. If the Profile button is shown depressed, this is telling you that a sketch or sketch area needs to be selected. If there are multiple closed profiles, you will need to select the profile that you want to sweep. If there is only one possible profile, Inventor will select it for you and you can skip this step. If the wrong profile or sketch area is selected, reselect the Profile button and choose the new profile or sketch area.

Path: Select this button to choose the path to sweep the profile along. The path can be either open or closed.

Taper: Type the angle that you want the sketch to be drafted. By default, the taper angle is 0 degrees.

OPERATION

This is the column of buttons along the right side of the dialog. By default the Join operation is selected. You can use the other operations to add or remove material from the part using the join or cut options, or keep what is common between the existing part and the completed sweep by using the intersect option. The surfaces button allows you to create a construction surface. Construction surfaces are discussed later in this chapter.

Join: Adds material to the part.

Cut: Removes material from the part.

Intersect: Keeps what is common to the part and the swept feature.

Surfaces: Sweeps the profile along the selected path and the result is a surface.

TUTORIAL 7.1—SWEEP FEATURES

In this tutorial you will create two sketches and then use the two sketches to create a sweep feature.

1. Start a new part file using the Standard.ipt template.
2. Sketch a line and an arc as shown in Figure 7.3. Project the center point from the origin folder and start the endpoint of the line from the projected center point.

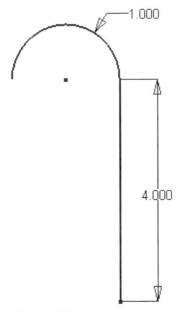

Figure 7.3

3. Click the Return tool.
4. Change to an isometric view.
5. Create a new sketch on the default XZ plane.
6. Sketch and dimension a .5-inch circle that is coincident with the end of the line. You will need to project the line onto the sketch, then add a coincident constraint between the center of the circle and the projected point.
7. Click the Return tool to exit the Sketch environment.
8. Issue the Sweep tool from the Features Panel Bar.
9. The Profile and Path are automatically selected for you. If you need to select either a path or profile, you can do so by selecting the appropriate button in the Sweep dialog box.
10. Click OK in the Sweep dialog box.
11. Your screen should look similar to the one shown in Figure 7.4.
12. Close the file without saving changes.

Figure 7.4

COIL FEATURES

Using the Coil feature you can easily create many types of helical or coil shapes. You can create many types of springs by selecting different settings in the Coil dialog box. You can also use the coil feature with the *cut* type of operation to cut a helical through the outside of a part if you want to represent threads by removing material from a part.

To create a coil, you will need to have one unconsumed sketch available in the part. This sketch will be the profile (or shape) of the coil. If no unconsumed sketch is available, Inventor will prompt you with an error message that states there are "No unconsumed visible sketches on the part." After an unconsumed sketch is available, you can select the Coil tool from either the Features Panel Bar, as shown in Figure 7.5 or Features toolbar. The Coil dialog box will open as shown in Figure 7.6 and give you the following options.

COIL SHAPE TAB

Profile: Select this button to choose the sketch to use as the shape of the coil feature. By default, the Profile button is shown depressed; this is telling you that a sketch or sketch area needs to be selected. If there are multiple closed profiles, you will need to select the profile that you want to revolve. If there is only one possible profile, Inventor will select it for you and you can skip this step. If the wrong profile

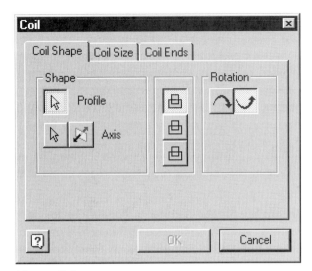

Figure 7.5   **Figure 7.6**

or sketch area is selected, reselect the Profile button and choose the new profile or sketch area. Only one closed profile can be used to create the coil feature.

Axis: Select this button to choose a straight edge or centerline that the sketch should be revolved about. The edge must be part of the sketch. If a sketched centerline is used as the axis, it also needs to be part of the sketch. The axis cannot intersect the profile.

Flip: Select this button to change the direction that the coil will be created from the axis. The direction will be changed in either the positive or negative x or y-axis depending upon the edge or axis that is selected. You will see a preview of the direction that the coil will be created.

Rotation: Specifies the direction that the coil will rotate. You can select to have the coil rotate either clockwise or counterclockwise.

OPERATION

This is the column of buttons along the center of the dialog that will appear if a base feature exists. The operation column is only available if the Coil feature is not the first feature in the part. By default the Join operation is selected. You can use the other operation to add or remove material from the part using the join or cut options, or keep what is common between the existing part and the completed coil feature by using the intersect option.

Join: Adds material to the part.

Cut: Removes material from the part.

Intersect: Keeps what is common to the part and the swept feature.

COIL SIZE TAB

The Coil Size tab, as shown in Figure 7.7, allows you to specify how the coil will be created. You are presented with various options for the type of coil that you want to create. Based on the type of coil that you select, the other parameters for Pitch, Height, Revolution, and Taper will become active or inactive. Specify two of the three parameters that are available and Inventor will calculate the last field for you.

Type: Select the parameters that you want to specify: Pitch and Revolution, Revolution and Height, Pitch and Height, or Spiral.

Pitch: Type in the value for the height that you want the helix to elevate with each revolution.

Revolution: Specify the number of revolutions for the coil. A coil cannot have zero revolutions. Fractions can be used in this field. For example, you can create a coil that contains 2.5 turns. If end conditions are specified (see next section), the end conditions are included in the number of revolutions.

Height: Specify the height of the coil. This height is measured from the center of the profile at the start to the center of the profile at the end.

Taper: Type the angle that you want the coil to be tapered. A Spiral coil type cannot be tapered.

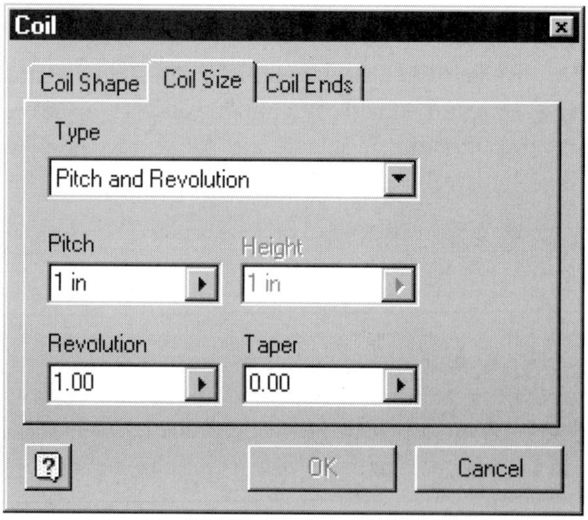

Figure 7.7

COIL ENDS TAB

The Coil Ends tab, as shown in Figure 7.8, lets you specify the end conditions for the Start and End of the coil. When selecting the Flat option, the helix is flattened, not the profile that you selected for the coil. The ends can have end conditions that are not consistent between the start and end of the coil.

Start: Select either Natural or Flat for the start of the helix. Click the down arrow to change between the two options.

End: Select either Natural or Flat for the end of the helix. Click the down arrow to change between the two options.

Transition Angle: This is the distance (specified in degrees) that the coil achieves the transition. Normally occurs in less than one revolution.

Flat Angle: This is the distance (specified in degrees) that the coil extends after transition that does not have a pitch (flat). It provides the transition from the end of the revolved profile into a flattened end.

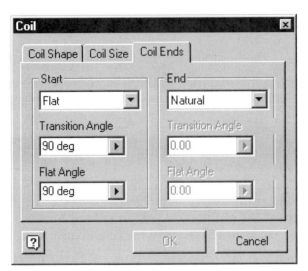

Figure 7.8

TUTORIAL 7.2—COIL FEATURES

In this tutorial you will create a torsion spring. You will create a coil feature that represents the spring and add a sweep feature that represents the ends of the spring.

1. Create a new part files based on the *Metric* Standard (mm).ipt template.
2. Create a sketch consisting of a circle (profile of the coil). Constrain the profile to be 10 mm below the X-axis and placed on the Y-axis. The diameter of the circle should be 3.43 mm. The sketch should look like Figure 7.9.

Note: Project the X-axis and Y-axis from the Origin folder to constrain the sketch geometry. The X-axis will be used as the axis of revolution for the coil.

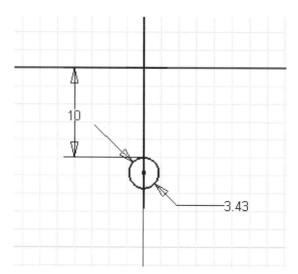

Figure 7.9

3. Click the Return tool to exit the sketch environment.
4. Change to an isometric view.
5. Click the Coil tool from the Features Panel Bar.
6. Because there is only one closed profile in the part, the Circle will automatically be selected as the Profile.
7. Select the projected X-axis as the axis of revolution.
8. Click the Coil Size tab.
9. Specify Revolution and Height from the Type drop-down list.
10. Enter a height of 23.5 mm, and 6.89 as the number of Revolutions as shown in Figure 7.10.
11. Verify that Natural is selected for both the Start and End of the coil from the Coil Ends tab of the Coil dialog box.
12. Click OK to create the coil feature.
13. Examine the feature and return to an Isometric View.
14. The coil should look similar to Figure 7.11.
15. Create a new sketch on the end of the coil that is furthest from you as shown in Figure 7.12.
16. The edge of the circle is automatically projected and will be used as the profile of the sweep feature.

Advanced Modeling Techniques 305

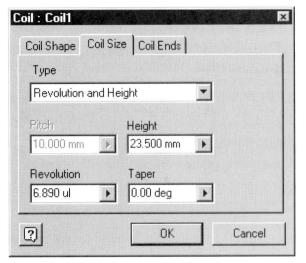

Figure 7.10

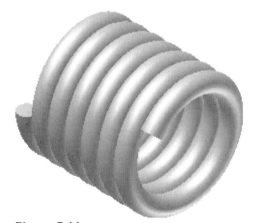

Figure 7.11

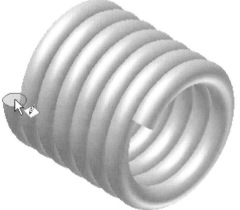

Figure 7.12

Note: Depending on your application settings, the circle may not be automatically projected. If it isn't, use the Project Geometry tool from the Sketch Panel Bar to project the circular edge of the end of the coil.

17. Click the Return tool.
18. Turn the visibility of the default XZ plane on and then create an offset work plane. The work plane should be parallel to the XZ plane and offset a distance of −10 mm. The work plane will be used to create a sketch representing the path for the sweep.
19. Create a new sketch on the work plane that you created.
20. Project the circle of the first sketch to the active sketch.
21. Sketch two lines and an arc and dimension them as shown in Figure 7.13. The vertical line must begin from the midpoint of the projected circle.

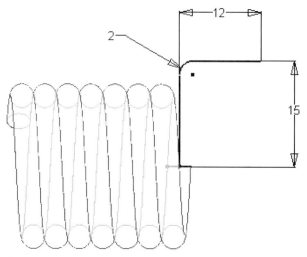

Figure 7.13

 Note: For clarity the visibility of the work plane has been turned off, the part is displayed in wireframe, and the view is shown in a plan view.

22. Click the Return tool to exit the Sketch environment. You will now have two unconsumed sketches that will make up the sweep feature.
23. Return to an Isometric View of the part.
24. Select the Sweep tool from the Features Panel bar.
25. If necessary, select the path of the profile for the sweep feature. Click OK to create the sweep feature.
26. When done, your part should look similar to Figure 7.14.

Advanced Modeling Techniques 307

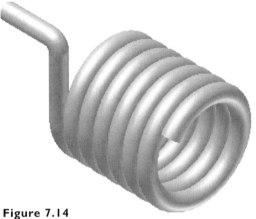

Figure 7.14

27. Figure 7.14
28. Create the other end of the torsion spring in a similar way that you just completed. Turn the visibility of the default YZ plane on and then create an offset work plane. The work plane should be parallel to the YZ plane and offset a distance of 23.5 mm.
29. Create a sketch on the flat end of the spring that will be the profile to be swept.
30. Create a new sketch on the work plane that you just created.
31. Click the Return tool to create an unconsumed sketch.
32. Project the circle to the active sketch.
33. Sketch a line and arc using the same dimensions shown in Figure 7.13. The sketch should look similar to Figure 7.15.

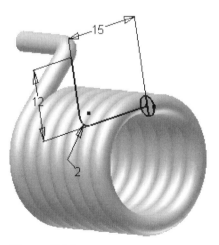

Figure 7.15

34. Click the Return tool to create the second unconsumed sketch.
35. Click the Sweep tool from the Features Panel Bar.
36. Select the Path and Profile if necessary.
37. Click OK to create the Sweep Feature.
38. The completed spring should look similar to Figure 7.16.
39. Save the file as \Inventor Book\ Spring.ipt

Figure 7.16

LOFT FEATURES

The Loft tool gives you the ability to create a feature that blends multiple profiles that may have different shapes. You can create complex shapes that contain two or more cross section profiles that reside on different planes. The profiles must be closed but they do not have to be parallel to each other. The loft feature will blend the shape of the profiles between the different work planes or existing part faces.

Loft features are used frequently in the creation of plastic or molded parts. Many of these types of parts have complex shapes that would be difficult to create using standard modeling features. There is no limit to the number of profiles that can be included in your loft feature and you also have the ability to have the feature start and end against edges of your part. To create a Loft feature, you select the Loft tool from the Features Panel Bar, as shown in Figure 7.17, or from the Features toolbar. The Loft dialog box will open as shown in Figure 7.18.

The Loft dialog box contains the following sections.

SECTIONS

This is where you will select the closed profiles (two or more) that will make up the lofted feature. The profiles can be created on any plane and do not have to be perpendicular to each other.

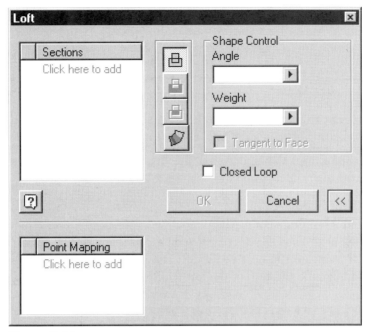

Figure 7.17   **Figure 7.18**

OPERATION

This is the column of buttons along the center of the dialog. By default the Join operation is selected. You can use the other operation to add or remove material from the part using the join or cut options, or keep what is common between the existing part and the completed loft by using the intersect option. You can create a construction surface from the loft operation using the Surface option. Surfaces will be covered later in this chapter.

Join: Adds material to the part.

Cut: Removes material from the part.

Intersect: Keeps what is common to the part and the loft feature.

Surfaces: Lofts the profiles and the result is a surface.

SHAPE CONTROL

You can control the angle, weight, and tangency of the loft feature using the edit boxes located in the Shape Control section of the Loft dialog.

Angle: Default is set to 90 degrees. The angle control sets the value for an angle formed between the plane that the profile is on and the direction to the next cross section of the loft feature.

Weight: Default is set to 0. The weight value controls the tangency of the loft shape to the normal of the starting and ending profile. A small value will create an abrupt transition and a large value creates a gradual transition.

Note: High values could result in twisting the loft and may cause a self-intersecting shape.

Tangent to Face: When checked, the lofted shape will remain tangent to the next terminating cross section of the loft feature.

Closed Loop: When checked, the first and last sections of the loft feature will be joined to create a closed loop.

MORE OPTIONS

Point Mapping: Point Mapping allows you specify *control* points from one section to the next. You can select a point on each section that you want to align with a specified point on the next profile. By using the point mapping functionality, you can minimize the twisting of the loft feature.

TUTORIAL 7.3—LOFT FEATURES

In this tutorial you will create a loft feature that transitions from a circular shape to a rectangle.

1. Create a new part file using the Standard.ipt template file.
2. Sketch a 2-inch diameter circle.
3. Click the Return tool to exit the sketch environment.
4. Change to an Isometric View of the part.
5. Create a work plane that is planar to the XY origin plane and offset a distance of 3 inches. Create a sketch on the new work plane.
6. Project the center of the sketched circle to the active sketch.
7. Create a rectangle that is centered about the projected point. The square should be 2.5 × 3 inches. Add dimensions to center the rectangle about the circle as shown in Figure 7.19.

Note: For clarity, the image is shown with the visibility of the work plane turned off.

8. Click the Return tool to exit the sketch environment and create the third unconsumed sketch in the part file.
9. Create another offset work plane a distance of 2 inches from the previous work plane that you created.
10. Sketch a 1.5-inch square. Use the same technique as before to make the square centered inside of the rectangle. Your part should look similar to Figure 7.20.

Advanced Modeling Techniques 311

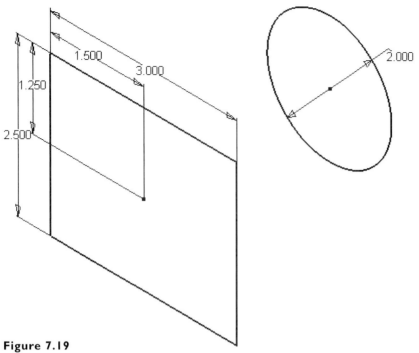

Figure 7.19

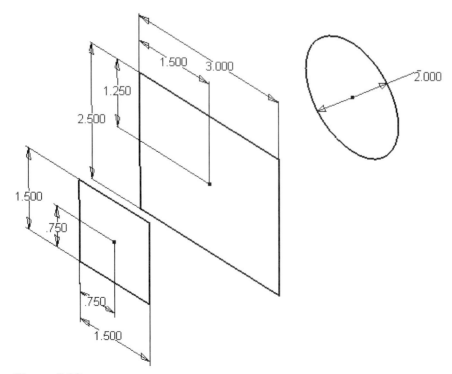

Figure 7.20

Note: The visibility of all work planes has been turned off for clarity.

11. Click the Loft tool from the Features Panel Bar.
12. Click on the text "Click Here to Add" from the Sections portion of the Loft dialog box.
13. Select the three profiles starting with the circle, then the rectangle and the square.
14. Click OK to create the Loft feature. The part should look similar to Figure 7.21.

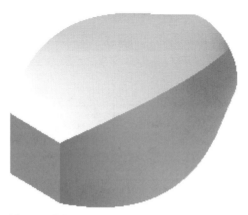

Figure 7.21

15. Experiment with the Angle and Weight control by editing the feature.
16. When finished, close the file without saving changes.

PART SPLIT

The split tool allows you to split parts using a sketched shape, surface, or a work plane. You can split an entire part by removing half of the part or split individual faces in order to apply draft to both halves of the part. You cannot use the split tool to create the other half of the part as is done in mold design. This can be accomplished using the Save As option and then modifying the split part feature to keep the other half of the part or by using Derived Parts (discussed later in the chapter). The Split tool is located on the Features Panel Bar as shown in Figure 7.22 or on the Feature toolbar. Once selected, the Split dialog box will be displayed as shown in Figure 7.23.

The Split dialog contains the following sections.

METHOD

Split Part: You can Split an entire part by selecting a work plane, surface, or sketched geometry; then use it to cut the part (removing material) in half. You are

Figure 7.22

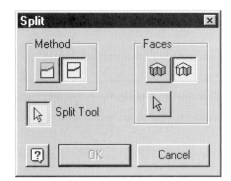
Figure 7.23

prompted for the direction of material that you want to remove. After splitting a part, you can retrieve the cut material by editing the split feature and selecting to remove the opposite side or by deleting the Split feature. The option to Remove material is only available when the Split Part method is used, as shown in Figure 7.24.

Split Face: You can Split individual faces of a part by selecting a work plane, surface, or sketched geometry and then selecting the faces that you want to split. The split feature can be edited to redefine the faces or to modify the feature to split the entire part instead of individual faces. The ability to select Faces is only available when the Split Faces option is selected as shown in Figure 7.23.

Split Tool: Select this button and then choose either a planar surface, work plane, or sketch that you want to use to split the part.

FACES

All: This button will select all faces of the part to be split.

Select: By choosing the Select button, you can choose specific faces that you want to split. After clicking the Select button, the Select Faces to Split tool becomes active.

Faces to Split: Select this button and then click on the faces of the part that you want to split. Any faces selected will be broken into two halves.

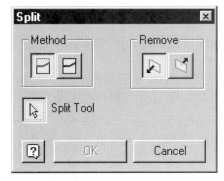
Figure 7.24

TUTORIAL 7.4—SPLITTING A PART

1. Open the file \Inventor Book\Split.ipt.
2. Create and dimension a new sketch on the front vertical face that contains an angled line as shown in Figure 7.25.

 Note: For clarity, the model is shown in wireframe display.

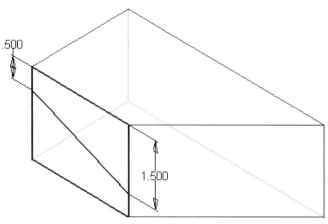

Figure 7.25

3. Issue the Split tool from the Features Panel Bar.
4. Select Split Part from the Method portion of the Split dialog box.
5. Click the Split Tool button and select the angled sketch line.
6. Select the Remove button that is not currently selected to see how Inventor previews the portion of material that will be removed. Your screen should look similar to Figure 7.26.
7. Click OK to split the part. When done, your part should resemble Figure 7.27.
8. Right click the Split feature in the Browser and select Edit Feature from the context menu.
9. Change the Method from Split Part to Split Face.
10. Select All from the Faces section of the dialog box.
11. Click OK to complete the operation and Split the faces of the part instead of removing material. When done, your part should resemble Figure 7.28.
12. Save the file.

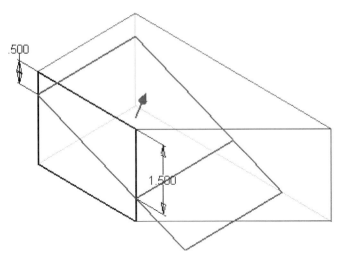

Figure 7.26

Figure 7.27

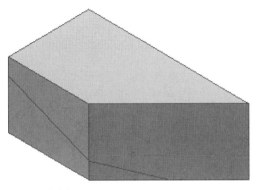

Figure 7.28

COPY FEATURES

You can copy and paste features between part files (both files must be open) or within a part file. When making a copy of a feature you have the option to designate if you want the feature, its dependent(s) and also parameters to be dependent or to be independent of the parent feature(s).

Creating a copy of a feature is similar to using iFeatures (discussed later in this chapter), but there are some differences:

- Only selected features are copied by default, not children.
- Upon pasting the feature, you can choose to copy dependent features.
- You can specify what plane or orientation you want the pasted feature to have when creating the pasted feature.
- The newly pasted feature and any dependents are completely independent and contain their own sketches and feature definitions within the Browser.

There are multiple ways of using the Copy/Paste feature within Inventor. After selecting a feature(s) in either the Browser or the graphics window:

- Initiate the context menu (right mouse button) and use the Copy and Paste options.
- Use the Copy and Paste selections from the Edit pull-down menu.
- Use the standard Window's shortcut key's CTRL+C & CTRL+V.

After selecting the Paste command using one of the methods mentioned above, the Paste Features dialog will open as shown in Figure 7.29.

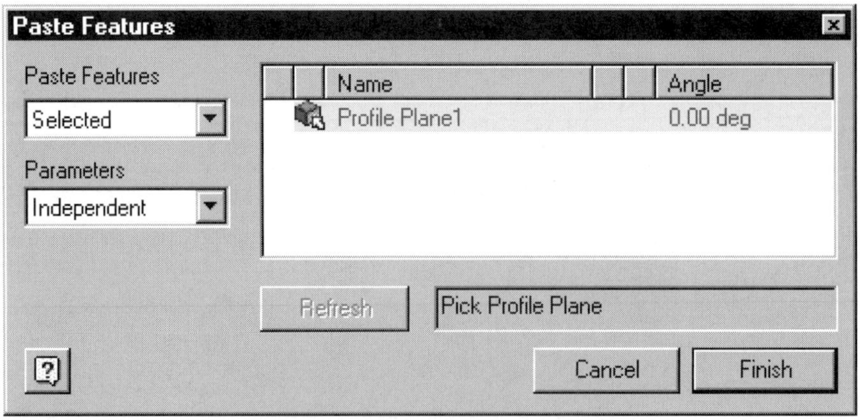

Figure 7.29

The Paste Feature dialog box contains the following fields:

Paste Features: Select to paste either only the Selected feature(s) or the selected feature(s) and its dependents from the drop down list.

> **Selected:** This will copy only the features that were selected when initiating the Copy command.
>
> **Dependent:** This will copy the dependent feature(s) of the selected feature(s).

Parameters: Select to make the parameters of the newly copied feature dependent on the parent feature or independent of the parent feature.

> **Dependent:** The dimensions of the copied feature(s) will be equal to the dimensions of the parent feature(s). For example, if the parent feature was defined with a value of d4=8 inches, the copied feature would have a parameter of d14=d4. The Dependent option is only available when copying features within the same part.
>
> **Independent:** The dimensions of the copied feature(s) will contain their own values. Initially the value will equal the value of the parent feature. For example, if the parent feature was defined with a value of d4=8 inches, the copied feature will have a parameter of d14=8 inches. The value of d14 can be modified independent from the value of d4.

Pick Sketch Plane: Once you have completed the selections for the features and parameters that are being pasted, you will select the plane for placement of the feature.

The placement plane can be any surface on the part or a work plane that has been defined. You can then use the compass in the preview image to roughly place the profile of the copied feature. Select the four-headed arrows to move the profile. Select the arc around the arrows to rotate the profile. You can flip the direction of the pasted feature by selecting the flip icon in the paste feature dialog box. The value for the angle of rotation of the profile can also be typed directly into the dialog box.

TUTORIAL 7.5—COPY / PASTE FEATURE

1. Open the file \Inventor Book\Copy Features.
2. Right click Extrusion2 in the Browser and select Copy from the context menu.
3. Right click in the graphics window and select Paste from the context menu. The Paste Features dialog box will open.
4. Change the Parameters drop-list to Dependent.

5. Select the angled face of the part.
6. Click the cross-hairs as shown in Figure 7.30. Then move the profile of the feature to the top right corner of the face as shown in Figure 7.31.

Note: Click once to select the cross-hairs and then click a second time to place the profile of the feature.

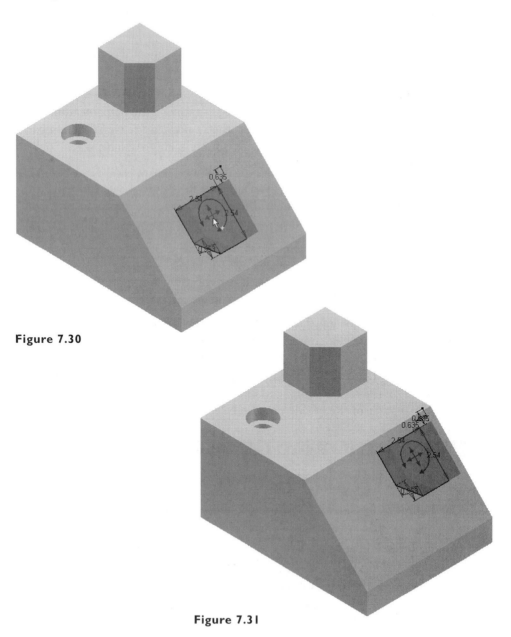

Figure 7.30

Figure 7.31

7. Click Finish in the Paste Features dialog box.

 Note: A new extrusion is added to the Browser representing the Pasted feature.

8. Edit the sketch of Extrusion2 by expanding the feature in the Browser, right click Sketch3 and select Edit Sketch from the context menu.

9. Modify the 1-inch dimension on the right side of the sketch to make it 1.75 inches as shown in Figure 7.32.

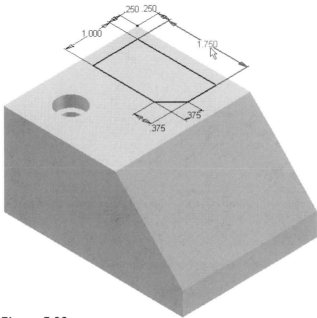

Figure 7.32

10. Click the Return tool.

 Note: The Pasted feature updates to match the parent extrusion because Dependent was selected for the Parameters in the Paste Feature dialog box. If Independent was chosen, the pasted feature would contain its own parameters.

11. Right click Sketch 4 (under Extrusion4) and select Edit Sketch from the context menu. Next you will precisely place the Pasted Feature on the part.

12. Add a coincident constraint between the point that lies at the top right of the sketch and the top right corner of the angled face as shown in Figure 7.33.

 Note: For clarity, the part is shown in wireframe.

13. Click the Return tool to update the part.

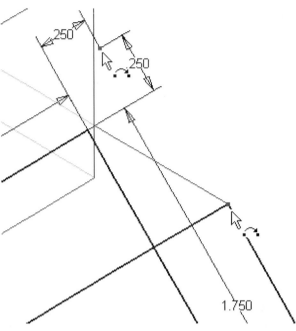

Figure 7.33

14. Add a .125 inch fillet to the top and bottom edges of the Extrusion on the top face of the part as shown in Figure 7.34.
15. Right click Extrusion2 in the Browser and select Copy from the context menu.

Figure 7.34

16. Right click in the graphics window and select Paste from the context menu.

17. Select Dependent from the Paste Features section and Independent from the Parameters section of the Paste Features dialog box as shown in Figure 7.35.

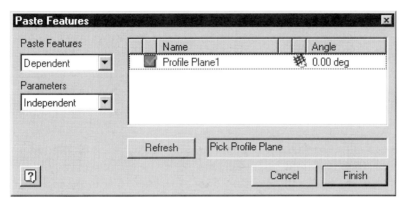

Figure 7.35

18. Select the front-left planar face as the Profile Plane and position the feature so that is will fit within the boundary of the face.

19. Click Finish in the Paste Features dialog box.

Note: The fillet was also copied because Dependent was selected from the Paste Features dialog. Because the parameters were made Independent, you can modify any of the parameters for the pasted feature independently of the parent feature.

20. The completed part should look similar to Figure 7.36.

21. Save and close the file. This part will be used in the next tutorial.

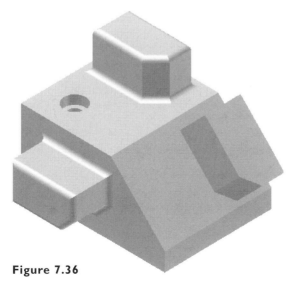

Figure 7.36

REORDER FEATURES

Features can be reordered within the Browser. To reorder features, select the feature that you want to move by clicking on the icon next to the feature in the Browser. Hold down the left mouse button (click-and-drag) and move the feature to the desired location in the Browser. A horizontal line will appear in the Browser to show you its relative location while reordering the feature, as shown in Figure 7.37. Release the mouse button and the remaining model features will be recalculated. If the feature cannot be moved due to parent-child relationships with other features, Inventor will not allow you to drag the feature to the new position and the cursor will change in the Browser to a *NO* symbol instead of the horizontal line as shown in Figure 7.38.

For example, if you've created a Fillet feature using the all fillets or all rounds option and then created an additional extruded feature (like a boss), you can move the fillet feature below the boss and have the edges of the new feature be included in the selected edges of the fillet.

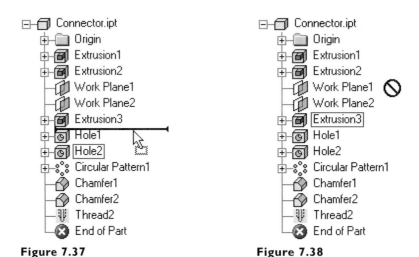

Figure 7.37   Figure 7.38

MIRROR FEATURES

You can mirror a created feature using the Mirror Feature tool. You mirror a feature by selecting the feature(s) that you want to mirror and then selecting a planar face on the part or a work plane to mirror the feature about. If the parent feature changes, the resulting mirror feature will also update. Select the Mirror Feature tool from either the Features Panel Bar, as shown in Figure 7.39, or from the Features toolbar.

Figure 7.39

The Mirror Pattern dialog will appear with the following options as shown in Figure 7.40.

Features: Select the feature(s) that you want to mirror.

Mirror Plane: Chose a planar face or work plane to mirror the feature(s) about.

Creation Method: Access the Creation Method tools by selecting the More button from the dialog.

Identical: Create mirrored features that remain identical regardless of where they intersect other features that are located on the part. Use this option when you want to mirror a large number of features.

Adjust to Model: Use this option when you want to calculate each feature that is being mirrored.

 Note: Use this option if the feature(s) being mirrored uses a model face as the termination option.

Figure 7.40

TUTORIAL 7.6—MIRRORING ABOUT A PLANE

1. Open the file \Inventor Book\Copy Features.ipt that you used in the previous tutorial.
2. Create a work plane that goes through the center of the part. There are a number of ways you can create this plane. One option is to use three points as shown in Figure 7.41.
3. Select the Mirror Feature tool.
4. Select the Extrusion4 and Fillet3 features from either the graphics window or the Browser.

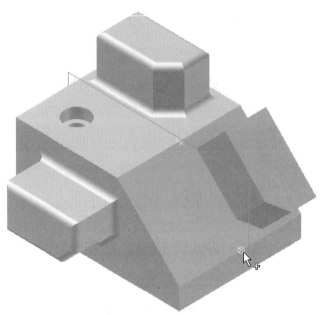

Figure 7.41

5. Select the Mirror Plane button from the Mirror Pattern dialog box.
6. Click the work plane that you created in step 2.
7. Click OK to create the mirror features and then close the Mirror Pattern dialog box. When done, your part should resemble Figure 7.42; for clarity the viewpoint has been rotated.
8. Save and close the file.

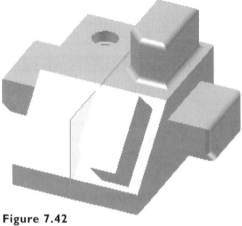

Figure 7.42

FEATURE SUPPRESSION

You can suppress features of a model to temporarily turn off the display of a feature(s). Features that are suppressed appear as a light gray color in the Browser and have a line drawn through them as shown in Figure 7.43. A suppressed feature will remain that way until it is unsuppressed, which will return the Browser display to its normal state. You can still use Inventor's highlight capabilities to see where the suppressed feature is located on the part. When your cursor is moved over the suppressed feature in the Browser, it will highlight in the graphics window.

To suppress a feature, select it from the Browser, then right click on the feature and select Suppress Features from the pop-up menu as shown in Figure 7.44. If the feature that you suppress is a parent to other features, those features will have the information icon next to them in the Browser. You can also select the feature in the graphics window and right click to access the Suppress Features option from the context menu. To unsuppress a feature, right click the suppressed feature in the Browser and select Unsuppress Features from the pop-up menu.

Feature suppression can be used to simplify parts, which also increases system performance. This capability also shows the part in different states through the manufacturing process and can be used to access faces and edges that otherwise you would not be able to reach. For example, if you wanted to dimension to the theoretical intersection of an edge that was filleted, you could suppress the fillet and add the dimension, then unsuppress the fillet feature.

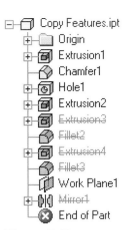

Figure 7.43

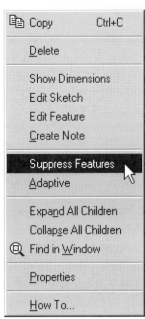

Figure 7.44

USING SURFACES

Surfaces can be created in Autodesk Inventor that can be used to help define the shape of your parts. Surfaces are created using the Extrude, Revolve, Sweep, or Loft tools by selecting the Surface icon from the Operation portion of the feature dialogs. The Surface option from the Extrude dialog is shown in Figure 7.45.

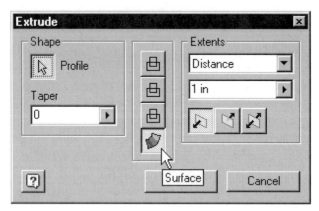

Figure 7.45

The surface data that is created can be used as either a split tool to divide part faces or remove a portion of a part, or as a termination face for other extrusion features. Surfaces are displayed as a translucent feature, similar to work features. Surface features are noted in the Browser with different icons and have the letters "Srf" appended to the feature name as shown in Figure 7.46.

Figure 7.46

The display of a surface feature can be toggled on or off using the context menu and selecting Visibility.

 Tip: You can use the Fillet and Chamfer tools to modify any sharp edges that are located on a surface.

RIBS AND WEBS

You can create ribs and webs, which are typically used as reinforcing or strengthening features in many mold and cast parts, using the Rib tool located on the Features Panel Bar, as shown in Figure 7.47, or from the Features toolbar. You must have an unconsumed sketch in order to create a rib feature. If there is no unconsumed sketch in the part file, Inventor will warn you with the message "No unconsumed visible sketches on the part."

Figure 7.47

A rib is a thin-walled feature that is typically closed. A web is similar in width but is usually an open shape. Rib networks can also be created when an entire series of thin-walled support features are needed. A rib or web feature is defined by a single open profile that is then refined using the options in the Rib dialog box shown in Figure 7.48.

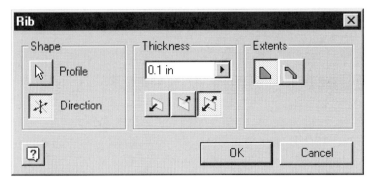

Figure 7.48

The Rib dialog box contains the following sections:

SHAPE
Profile: Select this button to choose the sketch to extrude. The sketch can be an open profile, or you can select multiple intersecting or nonintersecting profiles to define a rib or web network.

Direction: Select this button to choose the direction for the rib or web. You can preview the profile by moving the cursor around the edges of the profile.

THICKNESS
Edit Box: Specify the width of the rib feature using the edit box.

Flip Buttons: The Flip buttons specify which side of the profile to apply the thickness value or to add the same amount of material to both sides of the profile.

EXTENTS
To Next: This button will extend the ends of the open profile to the next available surface.

Finite: The Finite button enables an edit box for you to specify an offset distance from the sketch geometry. By enabling the Finite button and entering a distance, you can create a web feature.

Extend Profile: Using the Finite option also enables the extend profile checkbox. The checkbox specifies whether or not to extend the endpoints of the sketch to the next available surface or to leave the ends of the open profile as determined by the end of the rib feature. If checked, the ends are extended; if cleared, the ends *cap* at the end of the sketched profile.

RIB NETWORKS

Rib networks can also be created using the Rib tool. You can sketch a series of sketch objects within a single profile to create a rib network. The process of creation is the same for creating a single rib, except the thickness is applied to all objects within the profile.

TUTORIAL 7.7—CREATING RIBS AND WEBS

In this tutorial you will open an existing part and create a rib feature. You will then edit the rib feature and transform it into a web and delete the feature to create a rib network.

1. Open the file \Inventor Book\Rib.ipt
2. Create a new sketch on Work Plane 1 that is visible in the part.
3. Click the Look At tool and select the work plane.
4. Right click in the graphics window and select Slice Graphics from the context menu.
5. Zoom in to the top portion of the part and sketch a 3-point arc as shown in Figure 7.49.

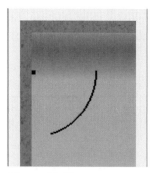

Figure 7.49

6. Click the Return tool to create an unconsumed sketch.
7. Change to an isometric view.
8. Select the Rib tool from the Features Panel Bar.
9. Specify the Direction to be in towards the part. The Rib previews as shown in Figure 7.50.

Advanced Modeling Techniques 329

Figure 7.50

10. Specify a Thickness of .125 inches using the Midplane option and the To Next termination from the Extents portion of the dialog.
11. Click OK to create the Rib.
12. Create a Rectangular Pattern of the Rib that contains 6 ribs with a .5 inch distance between them. When done your screen should resemble Figure 7.51.

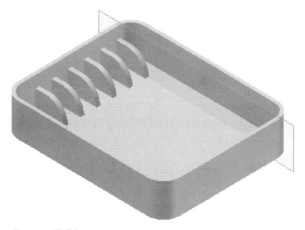

Figure 7.51

13. Right click on Rib1 in the Browser and select Edit Feature from the context menu.
14. Change the Extents to Finite and a distance of .125 inch.
15. Verify that Extend Profile is selected as shown in Figure 7.52.

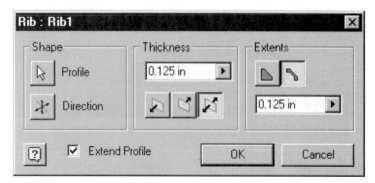

Figure 7.52

16. Click OK to complete the edit and modify the ribs to webs.
17. When done, your screen should resemble Figure 7.53.

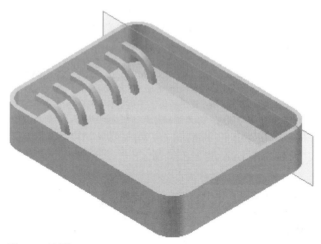

Figure 7.53

18. Right click on Rib 1 in the Browser and select Delete from the context menu, also delete Work Plane1.
19. Create a new sketch on the top surface of the part as shown in Figure 7.54.
20. Click the Look At tool and select the Sketch that was created in the Browser.
21. Sketch a series of lines as shown in Figure 7.55.

 Note: You can use the sketch pattern tool to define the series of lines. The lines do not need to be precise.

Advanced Modeling Techniques 331

Figure 7.54

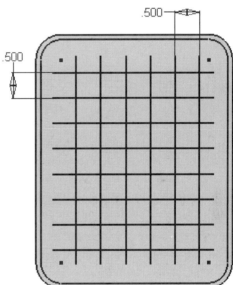

Figure 7.55

22. Switch to an Isometric View.
23. Select the Rib tool from the Features Panel Bar.
24. Choose all of the sketch lines as the profile.
25. Specify the direction to be towards the bottom of the part. Set the thickness to .125 inch, midplane option, and the Finite extent type and an extent of .125 inches.
26. Click OK to create the network.
27. When done your part should look similar to Figure 7.56.
28. Save the part.

Figure 7.56

iFEATURES

iFeatures give you the ability to reuse single or multiple features from a part file in other Inventor files. iFeatures capture the design intent that is built into the feature(s) such as the name of a feature as well as the size and position parameters for the feature that is going to be reused. You can also embed or attach a file to the iFeature that can be used as Placement Help when you place the iFeature into another design. After creating an iFeature, you can place it into another part by selecting a face to position the features stored in the iFeature.

Reference edges can be included as Position Geometry in your iFeatures. Reference edges allow you to capture additional design intent, but require that the iFeatures be positioned in the same way as it was originally designed in every part that the iFeature is placed.

iFeatures are saved in their own type of file that has an .ide extension.

CREATE iFEATURES

You create iFeatures using the Create iFeature tool that is available from the fly-out of the View Catalog icon on the Features Panel Bar or toolbar as shown in Figure 7.57. Once selected, the Create iFeature dialog box will be shown as in Figure 7.58.

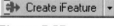

Figure 7.57

The Create iFeature dialog box contains the following sections.

Selected Features

This area of the dialog box lists the features that are selected to be included in the iFeature. Dependent features of a selected feature are included by default, but can be removed from the Selected Features list. Features in the list can be renamed to have a more descriptive name and assist in working with the iFeature at a later time.

The Add parameters (>>) button and the Remove parameters (<<) button allow you either to add or remove parameters from the highlighted features in the Selected Features list to the Size Parameters table.

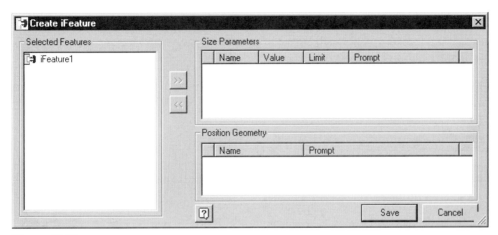

Figure 7.58

Size Parameters

The Size Parameters table lists all of the parameters that will be used for interface when the iFeature is placed in another file. You can select the parameters by expanding the features listed in the Selected Features list or directly from the graphics window.

 Note: Any parameters that have been given a name in the Parameters dialog box will automatically appear in the Size Parameters portion of the Create iFeature dialog box. Renaming the parameters that you want to include in your iFeatures in the Parameters dialog box can speed up the process of creation.

Name: Specify a descriptive name for the parameter. Use names that describe the purpose of the parameter. The name that is specified will appear in the browser when the iFeature is placed.

Value: The value placed will be the default for the parameter when inserting your iFeature into a file. The value is restricted by settings in the Limit column.

Limit: You can place restrictions on the values that are available for the parameter by using one of three options: None, Range, or List. *None* specifies that no restrictions are placed on the Value field. *Range* gives you the ability to specify a minimum and maximum value, including less than, equal to, and infinity symbols. You can specify the default value to be used. *List* allows you to define a list of values that the user can choose from for the size parameters.

Prompt: You can enter descriptive instructions to further explain what the parameter is used for. The text entered in the prompt field appears in a dialog box during the iFeature placement.

Position Geometry

Specify the geometry in the iFeature that is necessary to position it on a part. Geometry can be added or removed to the Position Geometry list from the Selected Features tree by right clicking the geometry and then selecting Add Geometry. You remove geometry from the Position Geometry list by right clicking the geometry to remove and selecting Remove Geometry.

Name: Describes the position geometry.

Prompt: Instruction for positioning the iFeature on a part.

TUTORIAL 7.8—CREATE iFEATURES

In this tutorial you will open an existing part file and create an iFeature from the part.

1. Open the file \Inventor Book\iFeature.ipt
2. Select the Parameters tool from the Standard Toolbar.

Note: Four parameters that are specific to the iFeature you will be creating have been renamed. These renamed parameters will be automatically added as Size Parameters of the iFeature.

3. Click Done in the Parameters dialog box.
4. Select the Create iFeature tool from the Features Panel Bar.

Note: The Create iFeature tool is located on the fly-out from the View Catalog icon.

5. Select Extrusion2 from the Browser. All dependent features of the feature will also be added as Selected Features as shown in Figure 7.59.

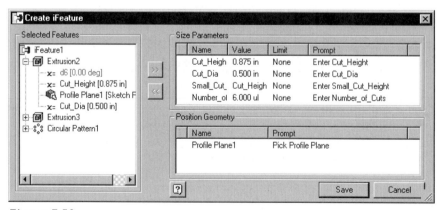

Figure 7.59

6. Select the Small Cut Height parameter and choose the Remove button (<<) to remove it as a size parameter.

 Note: This value is defined by an equation of the Cut Height, so it does not need to be included in the list.

7. Select the Limit column for the Cut Height parameter and choose Range from the provided list.

8. Specify a range of .25 to 1 with a default value of .875 as shown in Figure 7.60. Click OK to close the Specify Range dialog box.

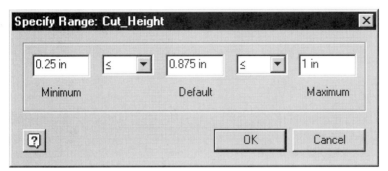

Figure 7.60

9. Specify a Range for the Cut Dia designating that the diameter must be between .25 and .5 inches as shown in Figure 7.61.

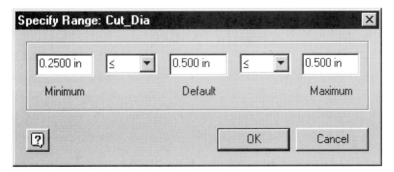

Figure 7.61

10. Specify a list of values for the Number of Cuts that includes 2, 4, 6, 8 and 10 cuts with 8 being the default value as shown in Figure 7.62.

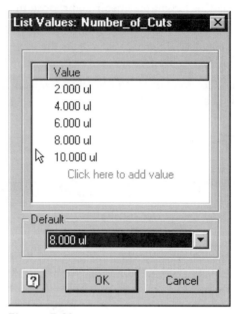

Figure 7.62

11. Click Save in the Create iFeature dialog and save the iFeature as Pattern Cuts in the \Inventor Book directory.

12. Close the iFeature.ipt file.

INSERT iFEATURES

You insert iFeatures using the Insert iFeature tool that is available from the fly-out of the View Catalog icon on the Features Panel Bar or toolbar as shown in Figure 7.63. Once selected the Insert iFeature dialog box will be shown as in Figure 7.64. You can select tasks from the tree structure on the left portion of the dialog or move forward and backward using the Next and Back buttons of the dialog.

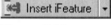

Figure 7.63

The Insert iFeature dialog contains the following sections.

Select: Choose the iFeature that you want to insert using the Browser button.

Position: Lists the names of the interface geometry specified during the creation process of the iFeature. After selecting a planar face or work plane, you can click the arrowhead on the positioning symbol to either move or rotate the symbol. You can also specify a precise rotation value directly in the angle field of the dialog box. After the requirement has been approved a check mark will be placed in the left column.

    Name: Lists the named interface geometry.

    Angle: Shows the default of the placement geometry on the iFeature.

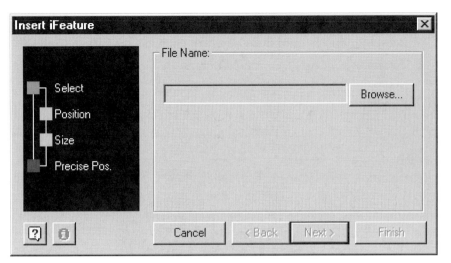

Figure 7.64

> **Move Coordinate System:** You can define horizontal or vertical axes when the iFeature has horizontal or vertical dimensions or constraints included.

Size: Shows the names and default values specified for the iFeature. Click in the row to edit the values and then Apply them to preview the changes.

> **Name:** Lists the name of the parameter.
>
> **Value:** Lists the value of the parameter.

Precise Position: After placing the iFeature, you can further position the iFeature using either dimensions or constraints.

> **Activate Sketch Edit Immediately:** Activates the sketch of the iFeature and the Sketch Panel Bar. You can then use dimensions and/or constraints to position the iFeature on the parent feature.
>
> **Do Not Activate Sketch Edit:** The iFeature is positioned without adding further constraints or dimensions.

TUTORIAL 7.9—INSERTING iFEATURES

In this tutorial you will create a part file and insert the iFeature that was created in the previous exercise.

1. Create a new part file based on the Standard.ipt template.
2. Sketch and Extrude a 2 inch cube.

3. Select the Insert iFeature tool from the Features Panel Bar.
4. Browse to the \Inventor Book directory and select the Pattern Cuts.ide file.
5. Select the front-right face as Profile Plane 1 as shown in Figure 7.65.

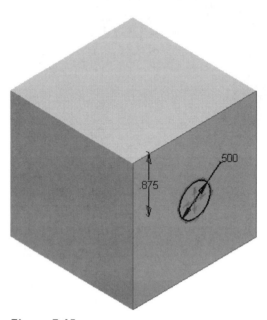

Figure 7.65

6. Click the position arrows and move the iFeature to the bottom left corner of the face.
7. Click Next.
8. Change the Cut Height to 1 inch.
9. Change the Cut Dia to .75 inch. A warning dialog box will be displayed stating that the value is outside the specified range.
10. Click OK in the warning dialog and leave the value at its default value of .5.
11. Change the Number of Cuts to 10.
12. Click Next.
13. Click Finish and do not activate the sketch for edit. Your screen should resemble Figure 7.66.
14. Experiment placing additional occurrences of the same iFeature in the part using different Size Parameters.
15. Close the file without saving changes.

Figure 7.66

iFEATURES OPTIONS

iFeatures are stored in a *Catalog*, the Catalog is simply a directory that is specified by you. You can set the location for the catalog and additional iFeature Options from the Application Options selection on the Tools pull-down menu. When the Options dialog box opens, select the iFeature tab as shown in Figure 7.67.

The iFeature tab of the Options dialog box contains the following settings.

iFeature Viewer: Specifies the viewer used to work with iFeature files. The default is Windows Explorer. You can enter the name of the executable file for the viewer application in the edit box.

iFeature Viewer Argument String: Sets a command line arguments for the viewer application. The default is /e which designates that Windows Explorer opens with the folder specified in the iFeature Root edit box.

iFeatures Root: Specifies the location of iFeature files used by the View Catalog dialog box. The location can be on your local computer or on a network drive. You can directly enter the path in the edit box or click Browse to select the path. The default path is the path to the Catalog folder installed with Autodesk Inventor.

iFeatures User Root: Specifies the location of iFeature files used by both the Create iFeature and Insert iFeature dialog boxes. The location can be on your local computer or on a network drive.

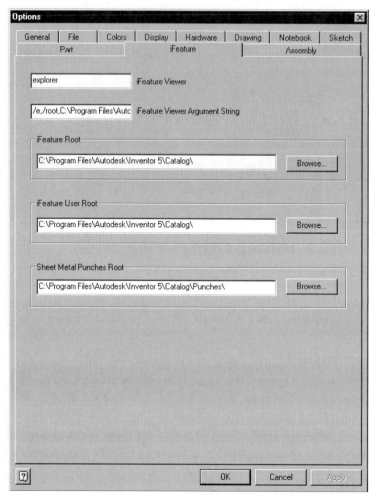

Figure 7.67

 Note: Windows shortcuts can be placed in either the iFeatures Root or the iFeatures User Root directories to quickly access other folders that contain iFeatures.

Sheet Metal Punches Root: Specifies the location of iFeature files used by the sheet metal Punch Tool dialog box. Sheet Metal Punches are discussed in Chapter 8.

DERIVED PARTS

Derived parts are used to capture design intent of an entire part, assembly, specific sketches, use parameters, or surfaces by either creating a new part file or *importing* any of the above selections. A derived part is a new part file that uses the selected part, assembly, or sketch as its base feature. Additional features can then be added to the derived part to explore different design scenarios. If the original part changes, the derived part will also update to include any changes that are made to the original file. This capability can also be turned off if you do not need to keep the tie to the original file.

The following are some possible uses for a derived part or assembly.

- You can explore design alternatives of a machined part from a casting blank.
- You can create a weldment so that assembly operations such as cuts or holes can be performed on the weldment.
- You can derive sketches to be used within an assembly as a layout.
- You can scale or mirror a derived part upon creation (not derived assemblies).
- You can derive a part as a work body that can be used to simplify the representation of a part.
- You can derive parameters from an existing file to be used within another part file.

You create a derived part using the Derived Component tool located on the Features Panel Bar as shown in Figure 7.68, or from the Features toolbar. You then select the Part or Assembly file that you want derived. If a part file is selected, the Derived Part dialog opens as shown in Figure 7.70. When an assembly file is selected as the Derived Component, the Derived Assembly dialog opens as shown in Figure 7.72.

Figure 7.68

 Note: When the Derived Component tool is not the base feature of the part, you have different options available in the Derived Part dialog box. For example, the ability to Derive a Solid Body or a Body as Work Surface cannot be selected. You cannot derive an assembly file unless it is derived as the base feature of a part.

DERIVED PART SYMBOLS

The symbols show you the current status of geometry that will be contained in the derived part and whether it will be included or excluded from the derived part. The symbols are shown in Figure 7.69.

Figure 7.69

The symbols in the Derived Part dialog have the following definitions:

Plus: Indicates that the selected geometry will be included in the derived part.

Slash: Indicates that the selected geometry will not be included in the derived part. Changes to this type of geometry in the parent part will not be incorporated or updated in the derived part.

Combination: Indicates that both included and excluded geometry types are contained in the derived part.

DERIVED PART DIALOG

The Derived Part dialog contains the following options as shown in Figure 7.70.

Solid Body: Select to derive the part as a base solid.

Body as Work Surface: Select to derive the part body as a work surface. The body is brought in, behaves, and appears as other Inventor surfaces.

Sketches: Select to include any sketches in the original file.

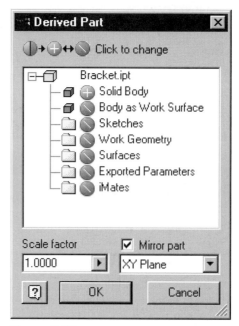

Figure 7.70

Work Geometry: Select to include any work features in the original file. They can then be used to create new geometry.

Surfaces: Select to include any surfaces that exist in the original file.

 Note: If the original file consists of only surfaces, it will be the only option available.

Exported Parameters: Select to include any parameters that are designated to be *exported parameters* in the original file.

 Note: You designate a parameter to be exported using the check box in the Parameters dialog accessed using the Parameters tool from the Standard toolbar.

iMates: Select to include any iMates that are defined in the original part.

Scale Factor: Indicates a scale factor (in percentage) to scale the derived part. The default scale factor is 1.0 (or the same size as the original file).

Mirror Part: Check this box to indicate that the derived part should be mirrored about the XY, XZ, or YX origin work planes upon creation. The plane to mirror about can be selected from the drop list.

DERIVED ASSEMBLY SYMBOLS

The symbols available when deriving an assembly perform different operations than the symbols displayed when deriving a part file. The derived assembly symbols are shown in Figure 7.71.

Figure 7.71

Plus: Indicates that the selected component will be included in the derived part.

Minus: Indicates that the selected component will be subtracted from the derived part. If the subtracted component intersects another portion of an included part, the result will be a void or cavity in the derived part.

Slash: Indicates that the selected component will not be included in the derived part. Changes to this type of component in the parent assembly will not be incorporated or updated in the derived part.

Combination: Indicates that the selected item contains a mixture of the above options.

DERIVED ASSEMBLY DIALOG

The derived assembly dialog contains the following options as shown in Figure 7.72.

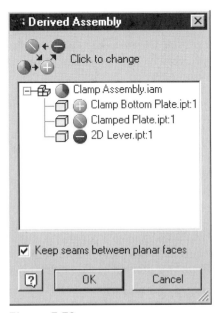

Figure 7.72

Click to Change: Select the symbol next to each component to designate it to be included or excluded from the derived part. You can also specify whether the component should be subtracted from the resultant derive part.

Keep Seams between Planar Faces: Select to create seams between any coincident planar faces of parts. If the box is cleared, the faces will be merged.

Because you are using geometry or files based on another file, any modifications to the parent will be incorporated into the derived component. If a parent file has been modified, the icon next to the derived component will show the Inventor update symbol as shown in Figure 7.73. You can break the link to the parent file by right clicking on the derived component in the Browser and selecting Break link from the context menu. Once the link is broken, the derived component icon will display a broken chain link to signify that the link to the parent file has been broken as shown in Figure 7.74. Once the link is broken, it cannot be relinked.

Figure 7.73   Figure 7.74

TUTORIAL 7.10—CREATE A DERIVED PART

In this tutorial you will derive a part from an existing file, mirror and scale the part, then update the parent file and see how the derived component updates.

Note: In this example you derived the body as Inventor surfaces. The surfaces can be used as termination faces for other features or to represent a simplified version of a part.

1. Create a new part file based on the Standard.ipt template.
2. Click the Return tool to exit the Sketch environment if it is active.
3. Click the Derived Component tool from the Features Panel Bar.
4. Select the Bracket.ipt file from the \Inventor Book directory. You worked on this file during an exercise in Chapter 2.
5. Select the Body as Work Surface option.
6. Enter a Scale Factor of .75.
7. Select the Mirror Part checkbox and mirror the part about the YZ plane. The completed dialog box should like as shown in Figure 7.75.
8. Click OK to create the derived part.
9. Return to an Isometric View of the part.

Advanced Modeling Techniques 345

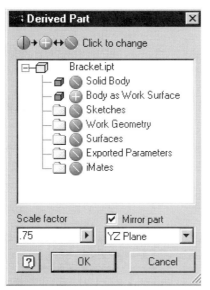

Figure 7.75

10. Right click Bracket.ipt (the derived component) in the Browser and select Edit Derived Part from the context menu.

11. Select Solid Body in the Derived Part dialog box and click OK.

12. Save the file as \Inventor Book\Derived Bracket.

13. Open the file \Inventor Book\Bracket that you created in Chapter 2–Exercise 2.1.

14. Add a .0625 inch fillet to the inside edges of both oblong cuts through the part and a .0125 inch fillet to the outer edge of the part as shown in Figure 7.76.

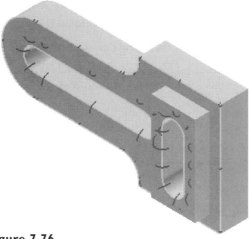

Figure 7.76

15. Save and close the Bracket.ipt file.
16. Notice the Update symbol has replaced the derived component symbol in the Browser of the part that contains the derived bracket, showing you that the parent component has been modified.
17. Select the Update tool from the Command Bar. The derived part updates to match the parent component and the derived part icon is restored in the Browser.
18. Right click Bracket.ipt in the Browser and select Break Link with Base Part.
19. The derived component icon in the Browser is now replaced with a broken chain link signifying that the link to the base part has been broken.
20. Save the file.

TUTORIAL 7.11—CREATE A DERIVED ASSEMBLY

In this tutorial you will create a derived assembly that represents a mold cavity by subtracting one part file from another.

1. Open the file Inventor Book\Mold.iam.
2. Examine the assembly of the two components. The knob has been constrained to be centered within the Base.
3. Close the Assembly.
4. Create a new part file based on the Standard.ipt template file.
5. Click the Return tool to exit the sketch environment.
6. Click the Derived Component tool from the Features Panel Bar.
7. Select the Inventor Book\Mold.iam file.
8. View the assembly as an Isometric View.
9. Click the name Knob in the dialog box until the *minus* symbol appears next to the component.
10. Select the option to *keep* seams between planar faces. The Derived Assembly dialog box should appear as shown in Figure 7.77.
11. Click OK to create the Derived Assembly.
12. Examine the contents of the Derived Assembly.
13. Close the file without saving changes.

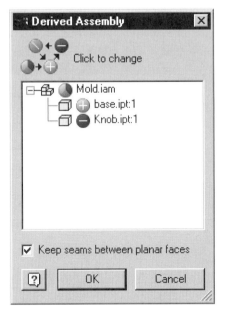

Figure 7.77

FILE PROPERTIES

You can specify properties for the files that you create in Autodesk Inventor using the Properties selection from the File pull-down. After selecting the Properties option, the Properties dialog box for the active file will be opened as shown in Figure 7.78.

The Properties dialog boxes for part and assembly files both contain six tabs, described below. The file properties for a drawing file contain the same tabs excluding the Physical tab. The properties of your files can be used to classify and manage your Inventor files. You can also use properties that have been populated with data as search criteria, to create reports, and to automatically update your title blocks and parts lists in drawings and bills of material. The tabs and the contents of the tabs are listed below.

SUMMARY
Contains fields for Title, Subject, Author, Manager, Company, Category, Keywords, and Comments.

PROJECT
Contains fields for Location, File Subtype, Part Number, Description, Revision Number, Project, Designer, Engineer, Authority, Cost Center, Estimated Cost, Creation Date, Vendor, and WEB Link.

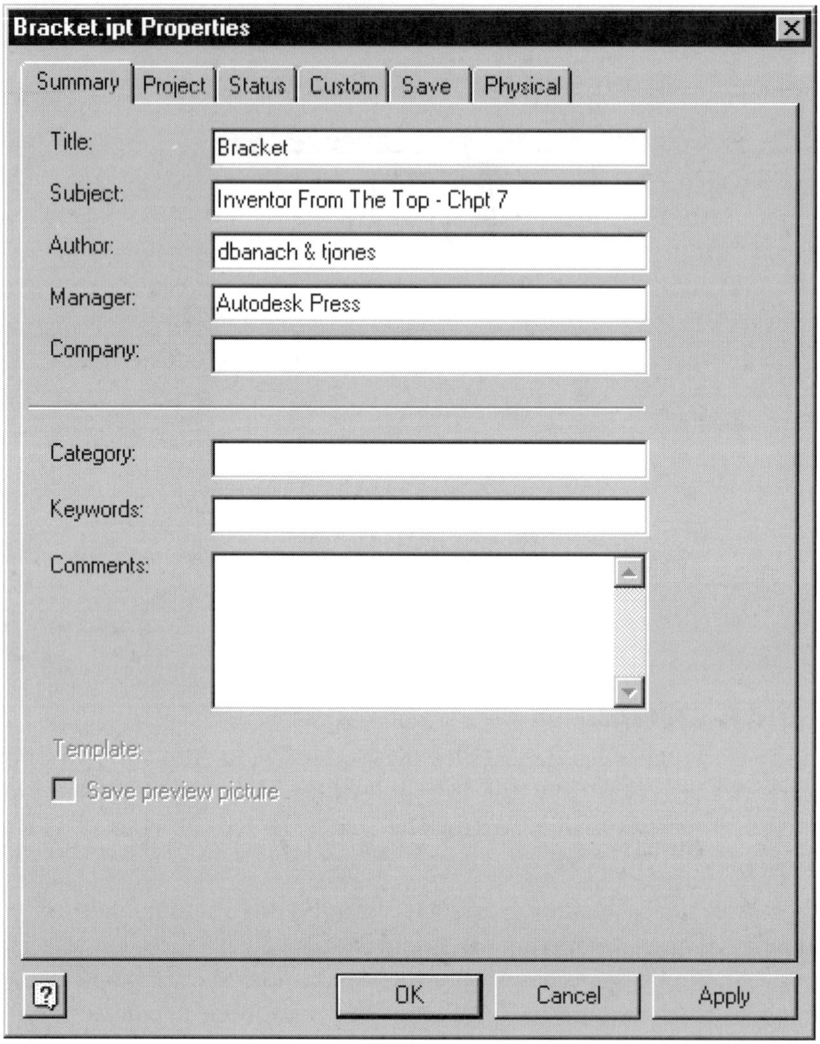

Figure 7.78

STATUS

Contains fields for Part Number, Status, Design State, Check By, Checked Date, Eng. Approved By, Eng. Approved Date, Mfg. Approved By, Mfg. Approved Date, File Status, Reserved By, Reserved, Last Reserved By, Reserve Removed.

CUSTOM

You can add custom properties to the active file using the Custom tab. The general Microsoft custom properties are also available and can be added to your Inventor files.

SAVE

You can specify whether you want to save a preview picture of your files and the parameters for the preview image or you can specify an image file to use as the preview picture.

PHYSICAL

The Physical tab calculates and displays the physical and inertial properties for a part or assembly file. The Mass, Surface Area, and Volume are automatically calculate. Physical properties of the model change as you create the structure of your files. You will need to select the Update button to update the physical properties based on changes to your models.

REVIEW QUESTIONS

1. Explain the process used to create a sweep feature.
2. When creating a coil feature there must be one unconsumed sketch available before using the Coil tool.   True or False?
3. A loft feature is created using only two unconsumed sketches.   True or False?
4. You can control the twisting of profiles by accessing a context menu after creating a loft feature.   True or False?
5. When splitting a part, both halves of a part are retained within the same file. True or False?
6. Features can be copied between parts using the Copy Feature tool from the Features Panel Bar or Toolbar.   True or False?
7. A rib or web feature is defined by sketching a closed profile.   True or False?
8. What is the benefit of iFeatures?
9. How do you create a mirrored part file?
10. Derived assemblies give you the ability to use parts as a Boolean operation from another part file.   True or False?

CHAPTER 8

Additional Functionality

This chapter will introduce you to a number of tools not covered in earlier chapters. You will use the sheet metal design environment to create part models that can be constructed by bending flat sheets using standard sheet metal fabrication tools. You will use the Engineers Notebook to add notes and other descriptive information to your models, and learn how to use the Design Assistant to track, manage, and report on your designs. You will also learn how to access the RedSpark parts library included with Inventor, and place standard components from the library. Finally, you'll examine the available methods for importing and exporting data in 2-D and 3-D formats, with an emphasis on AutoCAD files.

CHAPTER OBJECTIVES

After completing this chapter, you will be able to:

- use the Engineers Notebook to help document a design.
- set up a sheet metal style.
- create sheet metal faces, flanges, cuts, bends, and corner seams.
- create a flat pattern from a folded sheet metal part.
- export a flat pattern in DXF format.
- add standard components to an assembly from the RedSpark standard parts library.
- use Design Assistant to manage file properties, and make copies of designs.
- use the DWGOUT and DWGIN wizard to work with AutoCAD data.
- import 3-D files from other CAD systems.

ENGINEERS NOTEBOOK

In this section you will learn how to access the Engineers Notebook in the part and assembly environments, and learn how to add notes and other information to the notebook. Notes may be used to add information about a design feature, or to associate a design question to a specific component in an assembly. The Engineers Notebook enables you to attach additional information to sketches, features, parts, and assembly models. You might use this to save a record of how the design was built, or to add information about a component from outside sources, such as the results of an FEA analysis on a part.

ADDING NOTES TO A MODEL

To access the Engineers Notebook, select any sketch geometry, feature, part, or subassembly, then right click and select Create Note from the context menu. The Engineers Notebook appears in a separate window as shown in Figure 8.1. Each note initially consists of a screen capture of the current Inventor graphics window and a comment box for your note.

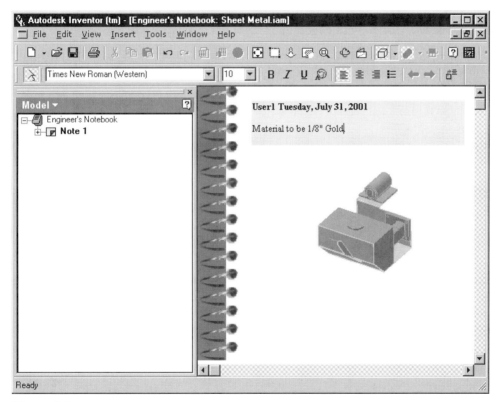

Figure 8.1

Following is a description of the tools available in the Engineers Notebook.

Text Creation

Basic text formatting tools are located in a toolbar at the top of the Notebook window. You can select from all available Windows fonts; change text height, color, style; and adjust text justification as you would in a word processor.

View Tools

The view below the text is a live window of the active model. View manipulation tools are available on the Standard toolbar or can be accessed by right clicking in the view. Use the standard function keys to pan (F2), zoom (F3), and rotate (F4) the view. Other view manipulation tools on the context menu include:

Delete: Deletes the view window. Additional views can be added to the notebook page by right clicking in a blank area of the page and selecting Insert View from the menu.

Display: Changes the display of solid models in the view. As in the main graphics window, shaded, shaded with edges, and wireframe display modes are available.

Freeze: Locks the content in the view window. This option enables you to capture the model at any design stage. A view of a feature sketch can be captured, or a history of design revisions can be stored in the notebook.

Note: If you uncheck Freeze, the view will update to match the current model view.

Tip: To eliminate the possibility of having the view change, use a screen capture program (or Ctrl+Print Screen) to capture a view of the screen and paste it into the notebook page as a raster image.

Restore Camera: Adjusts the view contents to match the current model window.

Arrow Tool

Use the arrow tool to draw attention to specific geometry in a view. Select Insert > Arrow from the application menu or right click in a blank area of the note and select Insert Arrow from the context menu. The arrow is placed at your first mouse click and leader segments are placed on subsequent mouse clicks. Right click and select Done to complete the leader. Move the complete arrow by clicking and dragging a leader segment. Modify the position of a leader vertex by clicking and dragging the vertex.

Comment Box and View Positioning

Add additional views and text comment boxes from the Insert menu or by right clicking in a blank area of the note and selecting Insert Comment or Insert View. Delete a comment box or view from the right click context menu for the object. Modify the position of comment boxes and views on the page by dragging them into place. Modify the size of a view or comment box by dragging a handle at the corner or midpoint of an edge. Arrow vertices inside a view or comment box retain their position when the view or comment is moved.

ADDING OTHER CONTENT

Additional data such as spreadsheets can be embedded or linked to a note using Windows Object Linking and Embedding (OLE). Select Insert > Object from the application menu or copy and paste objects into a comment box. Double click an embedded or linked object to edit it in its parent application.

MANAGING NOTES

The Engineers Notebook includes a Browser that lists the components of each note in the document. In an assembly, the Browser presents a hierarchical view of all components in the assembly and the notes associated with each component. Following is a list of the tools available from the Browser context menu. Right click in a blank area of the Browser to access the menu.

Arrange Notes: Notes can be sorted by Author Name, Date, Note Name (you can rename a note in the browser), or by the text in the first comment box.

Note: Each note is associated to a note author. The Username entered on the General tab of the Application options dialog box is the author name attached to a note. In a multiuser environment, notes from multiple authors may be present.

Insert Folder: Inserts a folder into the browser. Rename the folder to indicate its contents. You can drag and drop notes between Browser folders.

Place New Notes Here: Sets the default folder location for new notes added to the current document.

NOTE DISPLAY

Notes are represented in the graphics window by a small note symbol. When you hover over a note symbol, the text in the first comment box is displayed in a tool tip. Right click the note and select Edit (or double click the note) to open the Engineers Notebook. Notes are also listed in the part and assembly browsers, under the object that the note refers to. Settings on the Notebook tab of the Application Options dialog box enable the suppression of note symbols or tool tip text in the graphics window.

SHEET METAL PARTS

In this section, you will learn how to create sheet metal parts in Inventor. Assemblies often require components that are manufactured by bending flat metal stock to form brackets or enclosures. Cutouts, holes, and notches are cut or punched from the flat sheet and 3-D deformations such as dimples or louvers are often formed into the flat sheet. The punched sheet is then bent at specific locations using a press brake, or other forming tools, to create a finished part. The sheet metal environment in

Inventor presents specialized tools for creating models in both the folded and unfolded states.

Sheet metal parts can be designed separately, or in the context of an assembly. Since sheet metal parts are often used as supports for other components, or to enclose other components, designing in the assembly environment can be advantageous.

SHEET METAL DESIGN METHODS

Sheet metal parts are most often created in the folded state. The model is then unfolded into a flat sheet using the Flat Pattern tool. To create the folded model, the first sketch is extruded the thickness of the sheet to create a face. Additional faces or flanges are added to open edges of the part and bends are automatically added between the features. Cuts and other special sheet metal features are added to the part as required.

Sheet metal parts can also be created from standard parts that have been shelled. All walls of the shelled part must be the same thickness and match the thickness of the flat sheet. The solid corners of the shelled part are ripped open to enable the model to unfold, and appropriate bends are added along edges between faces.

Inventor's ability to create models with disjointed solids (unconnected features) is very helpful when building sheet metal parts in an assembly. Separated faces of a sheet metal part can be built quickly by referencing faces on other parts in the assembly. Additional sheet metal features can then join the distinct faces.

CREATING A SHEET METAL PART

The first step in creating a sheet metal part is to either select the New icon from the "What To Do" section, or select New from the pull down menu or the Standard toolbar, and then select the Sheet Metal.ipt icon from the Default template tab as shown in Figure 8.2. If the Standard.ipt file is used, the sheet metal tools can be loaded by picking Sheet Metal from the Applications pull down menu.

Sheet Metal.ipt

Figure 8.2

SHEET METAL TOOLS

After issuing the new sheet metal tool, Inventor's screen will change to reflect the sheet metal environment. As with standard parts, a sketch is created and is active. The sketch tools are common to both sheet metal and standard parts. The first sketch of a sheet metal part must either be a closed profile that will be extruded the sheet thickness to create a sheet metal face, or an open profile that will be thickened and extruded as a contour flange. Additional tools are used to add sheet metal features to the base feature. Following is a description of the tools used to create sheet metal features. You can also add standard part features to a sheet metal part, but these features may not unfold when a flat pattern is generated from the folded model.

SHEET METAL STYLES

The sheet metal specific parameters of a part are stored in a sheet metal style. These include the thickness and material of the sheet metal stock, a bend allowance factor to account for metal stretching during the creation of bends, and various parameters dealing with sheet metal bends and corners. You can create additional sheet metal styles to account for various materials and manufacturing processes or material types. If you create sheet metal styles in a template file, they will be available in all sheet metal parts based on that template.

The thickness of the sheet metal stock is the key parameter in a sheet metal part. Sheet metal tools such as face, flange, and cut automatically use the Thickness parameter to ensure that all features are the same wall thickness, a requirement for unfolding a model. In the default sheet metal style, all other sheet metal parameters such as bend radius are based on the Thickness value.

To edit or create a sheet metal style, follow these guidelines.

- Pick the Styles tool from the Sheet Metal Panel Bar as shown in Figure 8.3. The Sheet Metal Styles dialog box will appear as shown in Figure 8.4.

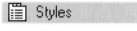

Figure 8.3

- To create a new sheet metal style, click the New button in the Sheet Metal Styles dialog box. A copy of the style currently displayed in the dialog box is created.

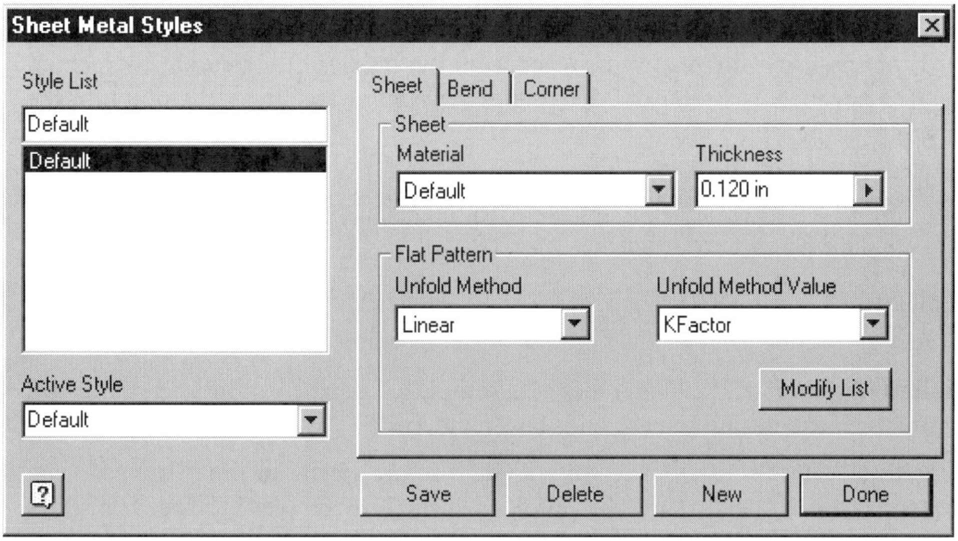

Figure 8.4

- Edit the values on the Sheet, Bend, and Corner tabs to define the feature properties of parts created with this sheet metal style.
- Rename the style and click the Save button.
- When multiple sheet metal styles exist in a part, set the active style by selecting it from the Active Style list.

Changing to a different sheet metal style, or making changes to the active style, updates the sheet metal part to match the new settings.

Following is a description of the settings in a sheet metal style. The parameters with numeric values, such as Thickness and Bend Radius, are saved as model parameters. The parameter values are updated when a sheet metal style is modified, or a new sheet metal style is activated. Other model parameters, and user defined parameters can reference these parameters in equations.

Sheet Tab (see Figure 8.4)
Material: Select from a list of defined materials in the document. The material color setting is applied to the part.

Thickness: Thickness of the flat stock that will be used to create the sheet metal part.

Unfold Method: Select the method used to calculate bend allowance. Bend allowance accounts for material stretching during bending. Options are a constant K Factor, or a bend table for more complex requirements.

Unfold Method Value: Combined with Unfold Method. The displayed value is the default value used for calculating bend allowances. A K Factor is a value between 0 and 1 that indicates the relative distance from the inside of the bend to the neutral axis of the bend. A K Factor of 0.5 specifies the neutral axis lies at the center of the material thickness.

Modify List button: Maintains a list of K Factors and named Bend Tables included in the sheet metal style. Each bend can have the default bend allowance overridden by a value in this list.

Bend Tab (see Figure 8.5)
Radius: The inside bend radius between adjacent, connected faces.

Minimum Remnant: If a bend relief cuts within this distance of the end of an edge, the small tab of remaining material is also removed.

Transition: Controls the intersection of edges across a bend in the flattened sheet. For bends without bend relief, the unfolded shape is a complex surface. Transition settings simplify the results straight lines or arcs, which can be cut in the flat sheet prior to bending.

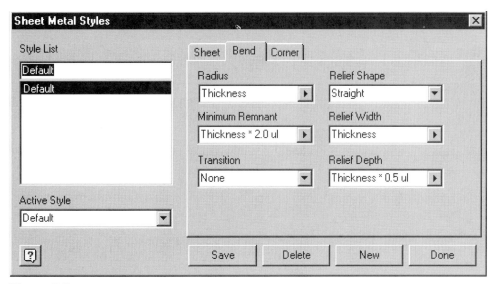

Figure 8.5

Relief Shape: If a bend does not extend the full width of an edge, a small notch is often cut next to the end of the edge to keep the metal from tearing at the edge of the bend. Square or round shapes are available.

Relief Width: Width of the bend relief.

Relief Depth: Setback amount from the edge. Round relief shapes require this be at least one half the Relief Width.

Corner Tab (see Figure 8.6)
A corner occurs where three faces meet. To enable efficient bending, a small amount of material is often removed from the flat sheet at the location where the three faces intersect.

Relief Shape: Specifies the shape of the corner relief.

Relief Size: Sets the size of the corner relief when the shape is set to round or square.

FACE

The Face tool extrudes a closed profile a distance equal to the sheet metal thickness. If the face is the first feature in a sheet metal part, you can only flip the direction of the extrusion. When a face is created later in the design, an adjacent face can be connected to the new face with a bend. If the sketch shares an edge with an existing feature, the bend is automatically added. You can optionally select a parallel edge on a disjointed face. This will extend or trim the attached face to meet the new face, with a bend created between the two faces. To create a sheet metal face, follow these guidelines.

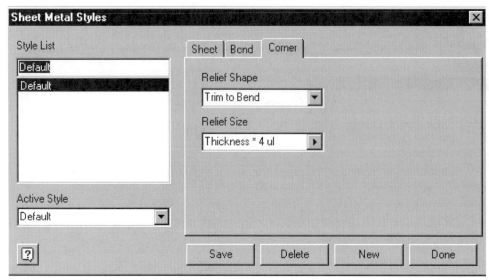

Figure 8.6

- Create a sketch with a single closed profile or a closed profile containing islands. The sketch is most often on a work plane created at a specific orientation to other part features or created by selecting a face on another part in an assembly.

 Note: You can create a single face from multiple closed profiles.

- Pick the Face tool from the Sheet Metal Panel Bar as shown in Figure 8.7. The Face dialog box will appear as shown in Figure 8.8.

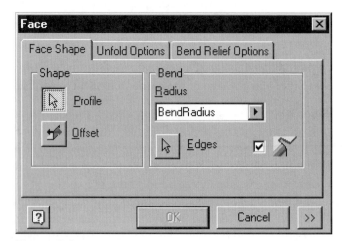

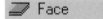

Figure 8.7   **Figure 8.8**

- If a single closed profile is available, it is selected. If multiple closed profiles are available, you must click inside one to define the face area.

- If required, flip the side to extrude the profile.

- If the face is not the first feature and the sketch does not share an edge with an existing feature, click the Edges button and select a parallel edge. The two faces are extended or trimmed as required, meeting at a bend. The bend is listed as a child of the new face in the Browser. If the face attached to the selected edge is parallel but not coplanar with the new face, a set of double bend options are presented. An additional face is added to connect the two parallel faces. The orientation and shape of this face is determined by the selected double bend option.

- Clear the checkmark from the Bend Relief Option if you don't want bend reliefs to be included.

- The Unfold Options and Bend Options tab contain settings for overriding the default values in the sheet metal style. Overrides are applied to individual features.

- Click OK.

CONTOUR FLANGE

A contour flange is created from an open sketch profile. The sketch is extruded perpendicular to the sketch plane and the open profile is thickened to match the sheet metal thickness. The profile does not require a sketched radius between line segments; bends are added at sharp intersections. Arc or spline segments are offset by the sheet metal thickness. A contour flange can be the first feature in a sheet metal part, or it can be added to existing features. To create a contour flange, follow these guidelines.

- Create a sketch with a single open profile. The sketch can contain line, arc, and spline segments. The sketch can be constrained to projected reference geometry to define a common edge between the contour flange and an existing face.

- Pick the Contour Flange tool from the Sheet Metal Panel Bar as shown in Figure 8.9. The Contour Flange dialog box will appear as shown in Figure 8.10.

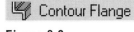

Figure 8.9

- Select the open sketch profile to define the shape of the contour flange.

- If required, flip the side to offset the profile.

- If the contour flange is the first feature in the part, enter an extrusion distance and direction. If the contour flange is not the first feature, an edge perpendicular to the sketch plane can be selected to define the extrusion extents. If the sketch is attached to an existing edge, select that edge. If the sketch is not attached to any projected geometry, the contour flange is extended or trimmed to meet the selected edge. The selected edge is typically the closest edge to the end of the sketch.

- Uncheck the Bend Relief checkbox to exclude bend reliefs between the contour flange and the selected edge.
- Apply any unfold and bend overrides on the Unfold Options and Bend Relief Options tabs.
- If an edge is selected to define the extrusion extents, you can select from three options to further refine the length of the contour flange. The following extent types can be specified.

    Edge: The contour flange extends the full length of the selected edge.

    Width: A point defines one extent of the contour flange. The flange can be offset from this point and extends a fixed distance. The starting point is defined by selecting an endpoint of the selected edge, a work point on a line defined by the edge, or a work plane perpendicular to the selected edge.

    Offset: Similar to the Width option, two points are selected that define both extents of the flange. An offset distance from each point can be specified.

- Click OK.

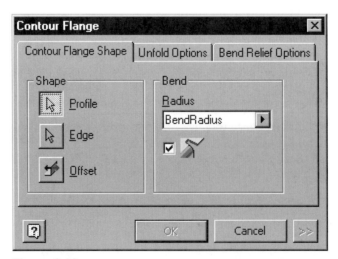

Figure 8.10

FLANGE

The flange tool creates a sheet metal face and a bend to an existing face. The selected edge defines the bend location between the two faces. To create a flange, follow these guidelines.

- At least one sheet metal face must already exist.
- Pick the Flange tool from the Sheet Metal Panel Bar as shown in Figure 8.11. The Flange dialog box will appear as shown in Figure 8.12.

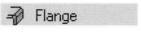

Figure 8.11

- Select an edge on an existing face.
- Enter the distance and angle for the flange in the Flange dialog box. The flange preview updates to match the current values.
- If required, flip the direction of offset side for the flange.
- Expand the Flange dialog box and select the appropriate extents type. See Contour Flange for additional information on extents.
- Apply the Flange to continue creating Flanges, or click OK to complete the Flange and exit the dialog box.

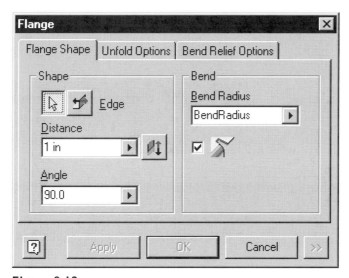

Figure 8.12

HEM

Hems are used to eliminate sharp edges or strengthen an open edge of a face. Material is folded back over the face with a small gap between the face and the hem. A hem does not change the length of the sheet metal part, the face is trimmed so the hem is tangent to the original length of the face.

To create a Hem, follow these guidelines.

- At least one sheet metal face must already exist.
- Pick the Hem tool from the Sheet Metal Panel Bar as shown in Figure 8.13. The Hem dialog box will appear as shown in Figure 8.14.
- Pick an open edge on a sheet metal face.
- Select the hem type.

    Single: A 180° flange.

    Double: A single hem is folded 180°, resulting in a double thickness hem.

    Teardrop: A single hem in a teardrop shape.

    Rolled: A cylindrical hem.

- Enter values for the hem. Teardrop and rolled hems require radius and angle values, single and double hems require a gap and length values. The hem preview changes to match the current values.
- Expand the dialog box and select Edge, Width, or Offset for the hem extents. See Contour Flange for additional information on defining extents.
- Apply the Hem to continue creating Hems, or click OK to complete the Hem and exit.

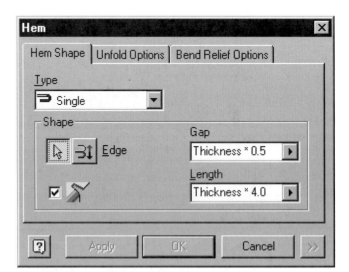

Figure 8.13

Figure 8.14

FOLD

An alternate method for creating sheet metal features is to start with a known flat pattern shape and add folds at sketched lines on a face. The fold tool can create sheet metal shapes that are difficult to create using face or flange tools. To create a Fold, follow these guidelines.

- Create a sketch on an existing face. Sketch a line between two open edges on the face. The line endpoints must be coincident to the face edge.
- Pick the Fold tool from the Sheet Metal Panel Bar as shown in Figure 8.15. The Fold dialog box will appear as shown in Figure 8.16.
- Pick the sketch line. The fold direction and angle are previewed in the graphics window. The fold arrows extend from the face that will remain fixed. The face on the other side of the bend line will fold around the bend line.
- If required, flip the fold direction and side.
- Enter the angle of the fold.
- Select the positioning of the fold with respect to the sketched line. The line can define the start, centerline, or end of the bend. The fold preview updates to match the current settings.
- Make any changes to Unfold or Bend Relief Options.
- Apply the Fold to continue creating Folds (an unconsumed sketch with valid geometry must be available), or click OK to complete the Fold and exit.

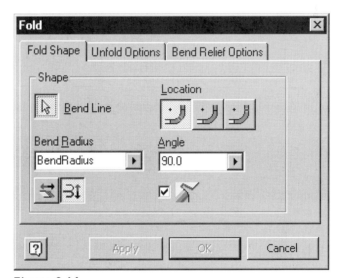

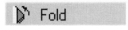

Figure 8.15   **Figure 8.16**

BEND

Bends are created as child objects of other features when the feature results in two faces being connected. Bends can also be created as independent objects between disjointed faces. To create a Bend, follow these guidelines.

- The sheet metal part must have two disjointed or intersecting faces. An example of intersecting faces is a shelled box that is being changed into a sheet metal part. The intersection of the box base and a wall is an edge that can be changed to a bend.
- Pick the Bend tool from the Sheet Metal Panel Bar as shown in Figure 8.17. The Bend dialog box will appear as shown in Figure 8.18.
- Pick the common edge of two intersecting faces, or pick two parallel edges on disjointed, non-coplanar faces. If the two faces are parallel, a set of Double bend options is presented. Depending on the position of the two faces, the double bend options will be either 45° and Fixed Edges, or 90° and Full Radius.
- Make any changes to Unfold or Bend Relief Options.
- Apply the Bend to continue creating Bends, or click OK to complete the Bend and exit.

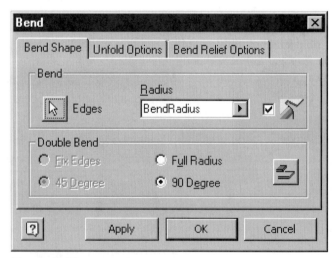

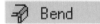

Figure 8.17   **Figure 8.18**

CORNER SEAM

When three faces meet in a sheet metal part, a gap is required between two of the faces to enable unfolding. Using a box as an example, the walls are connected to the floor, and gaps between the walls enable the box to be unfolded. The gap between adjacent faces is a corner seam. Corner seams can also be used to create mitered gaps between coplanar faces. Finally, a corner seam can be created by "ripping" open a corner at an intersection between faces. To create a Corner Seam, follow these guidelines.

- Pick the Corner Seam tool from the Sheet Metal Panel Bar as shown in Figure 8.19. The Corner Seam dialog box will appear as shown in Figure 8.20.
- Pick face edges that meet one of the following:

    Pick two open edges on faces that share a common connected edge. (Two walls that share a connection to a floor.)

    Pick two nonparallel edges on coplanar faces (creates a mitered joint with a gap between the extended faces.)

    Pick a single edge at the intersection of two faces. (The wall intersections of a shelled box.)

- Select the Miter option in the Corner Seam dialog box.
- Enter a value for the corner seam gap.
- Make any changes to the corner options.
- Apply the Corner Seam to continue creating Corner Seams, or click OK to complete the Corner Seam and exit.

 Note: The actual sheet metal thickness must equal the thickness in the sheet metal style for the Corner Seam operation to work.

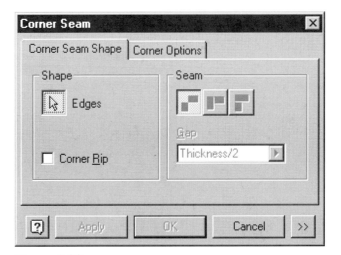

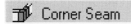

Figure 8.19

Figure 8.20

CUT

The Cut tool is a sheet metal specific implementation of the standard extrude tool. The Cut tool always performs an extrude cut; join and intersection are not available. The default extents are a blind cut, a distance equal to the sheet metal Thickness parameter. This ensures that the cut only extends through the face containing the sketch, and not through other faces that may be folded under the sketch face.

Most cuts are made on the flat sheet stock before the sheet is bent to form the folded part. The cuts often cross bend lines. Because the part is modeled in the folded state, representing cuts that cross bend boundaries requires the bend be unfolded to represent the flat sheet. When the bend is refolded, the cut deforms around the bend to ensure that the extrusion remains perpendicular to the sheet metal faces on both sides of the bend. When sketching a cut profile that will cross a bend, the Project Flat Pattern tool projects the unfolded flat pattern geometry onto the sketch face. To create a Cut feature, follow these guidelines.

- Create a sketch on a sheet metal face that includes one or more closed profiles representing the area(s) to cut. If required, use the Project Flat Pattern tool to project the unfolded geometry of connected faces onto the sketch.
- Pick the Cut tool from the Sheet Metal Panel Bar as shown in Figure 8.21. The Cut dialog box will appear as shown in Figure 8.22.
- Pick the profiles to Cut.
- If a profile crosses a bend, check the Cut Across Bend Option in the Cut dialog box.
- Apply the Cut to continue creating Cuts (an unconsumed sketch must be available), or click OK to complete the Cut and exit.

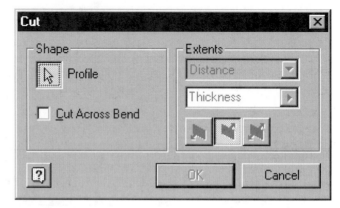

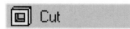

Figure 8.21   **Figure 8.22**

CORNER ROUND

The Corner Round tool is a sheet metal specific fillet tool. All edges, other than the ones at open corners of faces (all these edges have a length = Thickness), are filtered out. This enables you to easily select these small edges without zooming in on the part. To create a Corner Round, follow these guidelines.

- Pick the Corner Round tool from the Sheet Metal Panel Bar as shown in Figure 8.23. The Corner Round dialog box will appear as shown in Figure 8.24.
- Enter a radius for the corner round.
- Pick the Corner or Feature Select Mode in the Corner Round dialog box.
- Pick the corners or features to include.
- Add additional corner rounds with different radii, and pick corners or features for the additional corner rounds.
- Click OK.

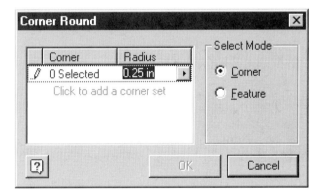

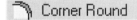

Figure 8.23

Figure 8.24

CORNER CHAMFER

The Corner Chamfer tool is a sheet metal specific chamfer tool. As with the Corner Round tool, all edges other than the ones at open corners of faces are filtered out. To create a Corner Chamfer, follow these guidelines.

- Pick the Corner Chamfer tool from the Sheet Metal Panel Bar as shown in Figure 8.25. The Corner Chamfer dialog box will appear as shown in Figure 8.26.
- Pick the chamfer style: One Distance, Distance and Angle, or Two Distances.
- Pick the corners (or edge and corner for Distance and Angle) to include.
- Enter chamfer values.
- Click OK.

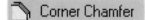

Figure 8.25

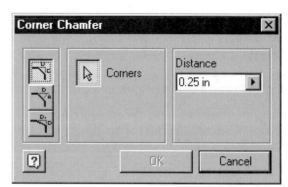

Figure 8.26

PUNCHTOOL

Cuts and 3-D deformation features such as dimples and louvers are usually added to the flat sheet metal stock in a turret punch prior to the sheet being bent into the folded part. Turret punch tools are positioned on the flat sheet by a center point corresponding to the tool center, and at an angle to a fixed coordinate system. The PunchTool places specially designed iFeatures on sketch points that define the center point of the tool. A streamlined interface simplifies the selection and placement of these iFeatures.

A default Punch folder is installed under the top level Catalog folder when Inventor is installed. A selection of punch tools is included in the folder. The PunchTool lists all iFeatures in the designated punch folder. You can set the default punch folder on the iFeature tab of the Application Options dialog.

To qualify as a PunchTool, a saved iFeature must have a single Hole Center point in the sketch of the first feature included in the iFeature. The point corresponds to the center of the tool. Nonsymmetrical shapes must have the center point position controlled by geometric relationships or equations to ensure that it remains centered when the iFeature parameters are changed. Create the iFeature in a sheet metal part to ensure that sheet metal parameters such as Thickness are saved with the iFeature. See the section on iFeatures in Chapter 7 for more information on creating iFeatures. To place a PunchTool, follow these guidelines.

- Create a sketch on a sheet metal face. Place at least one sketch point in the sketch. Hole Center points are automatically selected as punch centers during PunchTool placement, Sketch points, line or curve endpoints, and arc centers can be manually added as additional punch centers.

- Pick the PunchTool tool from the Sheet Metal Panel Bar as shown in Figure 8.27. The PunchTool dialog box will appear as shown in Figure 8.28.

- Select the desired PunchTool from the list in the PunchTool dialog box. Click Next.

- All Hole Centers are automatically selected. Hold down the SHIFT key and pick Hole Centers to exclude them, and pick other center points as required.
- Select a rotation angle for all occurrences of the punch. At zero degrees rotation, the X-axis of the first feature in the saved iFeature is aligned with the X-axis of the current sketch. Click Next.
- Enter values for the PunchTool parameters.
- Click OK.

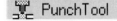

Figure 8.27

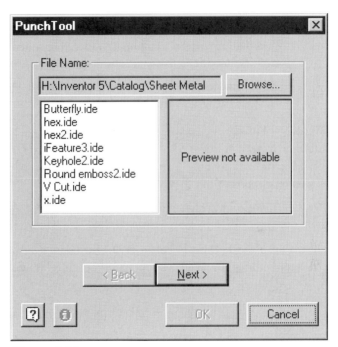

Figure 8.28

FLAT PATTERN

The final step is the creation of a sheet metal flat pattern that unfolds all the sheet metal features. The flat pattern represents the starting point for the manufacturing of the sheet metal part. The flat pattern can be displayed in a 2-D drawing view, complete with lines indicating bend centerlines. The Flat Pattern tool creates a 3-D model of the unfolded part. You can also directly export the flat pattern in 3-D (SAT file), or 2-D (DXF or DWG) formats, to be used to create machine tool programming for flat pattern punching or cutting. To create a Flat Pattern, follow these guidelines.

- Select a face. The selected face will remain fixed and all other faces will unfold from this face.

  Note: This is not a strict requirement, but selecting a face before creating a flat pattern is good practice.

- Pick the Flat Pattern tool from the Sheet Metal Panel Bar as shown in Figure 8.29. The Flat Pattern model is displayed in a new window. The flat pattern model is saved in the same document as the folded model.

  Figure 8.29

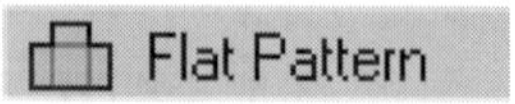

- Right click Flat Pattern at the top of the browser. There are menu items to save the flat pattern, align the flat pattern to a model edge, and display the overall dimensions (extents) of the flat pattern are available.

- To document a flat pattern, start a new drawing. Place a drawing view and select Flat Pattern from the Presentation View list in the Create View dialog box.

COMMON TOOLS

A number of tools on the Sheet Metal Panel Bar are common to sheet metal and standard parts. Following is a short description of how to use these tools in the sheet metal environment.

Work Features: Work planes are often used to define sketch planes for disjointed faces.

Holes: The Hole feature is common to both environments. You can enter sheet metal parameters such as Thickness in numeric fields such as hole depth.

Catalog Tools: iFeatures can be saved and placed in the sheet metal environment. See the PunchTool section for information on special iFeatures for sheet metal.

Mirror and Feature Patterns: Most sheet metal features cannot be mirrored or patterned. Holes, cuts, iFeatures, and features created with the standard part feature tools can be patterned or mirrored.

Promote: IGES files can be imported in the sheet metal environment.

TUTORIAL 8.1—SHEET METAL PART

In this tutorial you will open an existing assembly and create a sheet metal bracket using various sheet metal tools, to support the two parts in the assembly. You will attach a note to a part feature using the Engineers Notebook, and then create a flat pattern from the folded model. Finally, you will export the flat pattern in DXF format.

1. Open the file \Inventor Book\Sheet Metal.iam. The assembly should match the one shown in Figure 8.30.

Figure 8.30

2. Create a new sheet metal part in the assembly by picking the Create Component tool from the Assembly Panel Bar. Click the Browse button next to Template, pick the Sheet Metal (in).ipt template from the English tab, and click OK to accept the template. Enter SM03 in the New File Name edit box and ensure that Constrain sketch plane to face or plane is checked. Click OK and select the face highlighted in Figure 8.31.

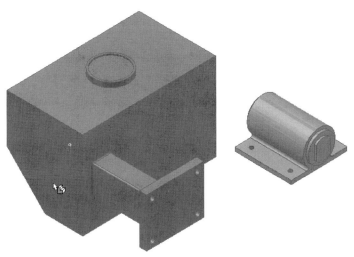

Figure 8.31

3. Pick the Project Geometry tool from the Sketch Panel Bar and click on the three edges of the large box highlighted in Figure 8.32. Avoid clicking the face of the box; the loop around the face will be projected if you do. Click the Return tool on the Standard Toolbar to return to the Sheet Metal environment.

4. Pick the Contour Flange tool from the Sheet Metal Panel Bar and select one of the three projected edges. In the Contour Flange dialog box, click the Offset tool and note that the preview of the Contour Flange now shows the material offset to the outside of the box. Enter 14 in the Distance edit box and click the Flip Direction button so the Contour Flange will extrude along the length of the box. Click OK, your view should resemble Figure 8.33.

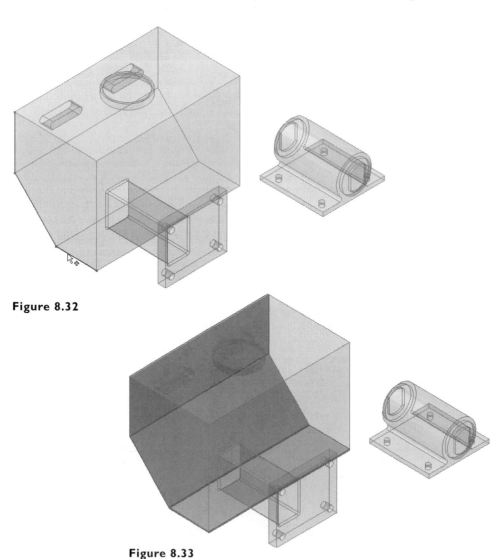

Figure 8.32

Figure 8.33

Additional Functionality 373

5. Create a sheet metal face flush to the rectangular mount on the near side of the box by picking the Sketch tool from the Standard toolbar and clicking on the face highlighted in Figure 8.34.

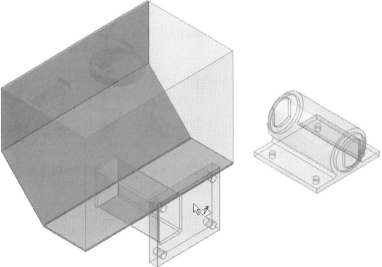

Figure 8.34

6. Pick the Project Geometry from the Sketch Panel Bar and click the two vertical edges highlighted in Figure 8.35. Pick the Two Point Rectangle from the Sketch Panel Bar and pick the upper endpoint of the project line on the left and the lower point of the projected line on the right. Then click the Return tool on the Standard toolbar.

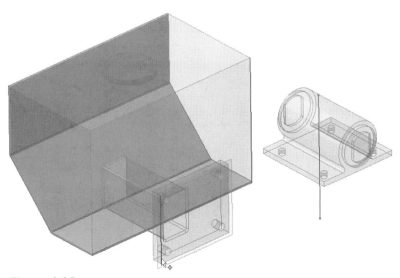

Figure 8.35

7. Click the Face tool on the Sheet Metal Panel Bar. Click the Edge highlighted in Figure 8.36 and then click OK to create the Face shown in Figure 8.37.

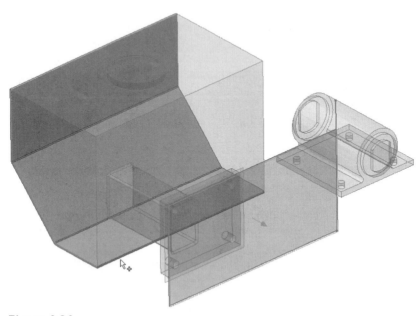

Figure 8.36

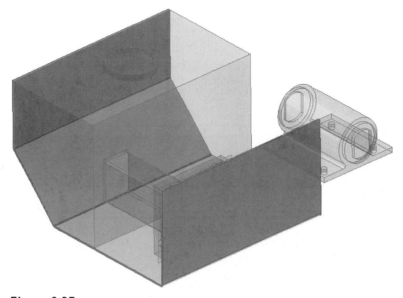

Figure 8.37

8. Join the two disjointed faces together by using the Bend tool. Start the bend tool and then pick the lower right edge of the bottom face and one of the lower edges of the right face.

9. Create a third face under the cylindrical part by selecting the Sketch tool from the Standard toolbar and picking the flat underside of the cylindrical part. Rotate the view or use the Select Other tool to pick the face. Project the same face (click inside the face to project all loops on the face) and create a sheet metal face and bend that is connected to the edge of the previous face. When complete, the sheet metal part should resemble Figure 8.38.

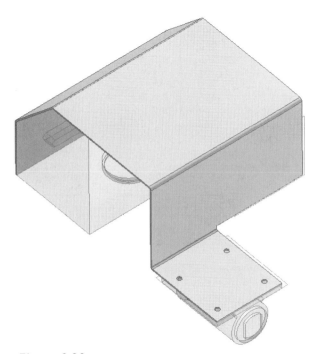

Figure 8.38

10. Clean up the view by turning off the visibility of the other parts in the assembly and the work planes in the sheet metal part. Reorient your view to resemble Figure 8.39.

11. Pick the Flange tool from the Sheet Metal Panel Bar, then click the edge highlighted in Figure 8.39. In the Flange dialog box, click the More button in the lower right corner to expand the dialog box, and pick Offset from the Extents Type list. Click the lower endpoint of the edge and then the upper endpoint of the edge. Enter **"0"** in the Offset1 edit box, and **"2"** in the Offset2 edit box. The preview should resemble Figure 8.40. Click Apply in the Flange dialog box.

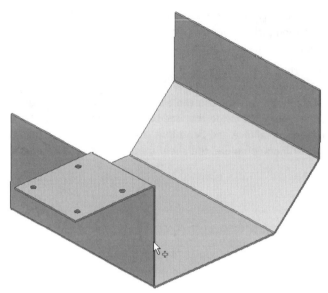

Figure 8.39

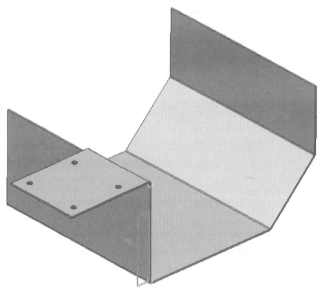

Figure 8.40

12. While still in the Flange dialog box, click the bottom adjacent edge next to the two holes to create a second flange. Click the left endpoint of the edge, and then the right endpoint of the edge, and click OK in the Flange dialog box. The second flange should match the one in Figure 8.41.

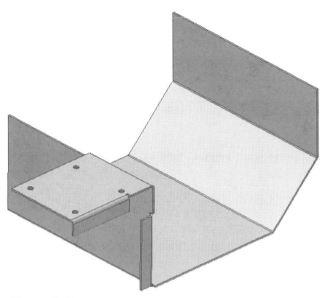

Figure 8.41

13. Create a Corner Seam to miter the two flanges by picking the Corner Seam tool from the Sheet Metal Panel Bar, then selecting the two edges highlighted in Figure 8.42. Enter Thickness/2 in the Gap edit box and click OK to create the Flange shown in Figure 8.42.

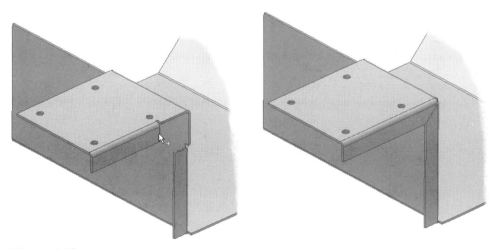

Figure 8.42

14. Create a Cut by picking the Sketch tool from the Standard Toolbar and then clicking on the face highlighted in Figure 8.43.

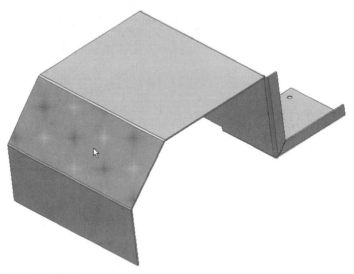

Figure 8.43

15. Click the arrow next to the Project Geometry tool in the Sketch Panel Bar, and pick the Project Flat pattern tool, then click the face highlighted in Figure 8.44.

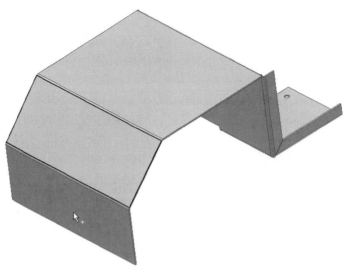

Figure 8.44

16. Use the Line tool to sketch a profile similar to the one in Figure 8.45 and then right click and pick Finish Sketch.

 Note: Dimensions can be added as required.

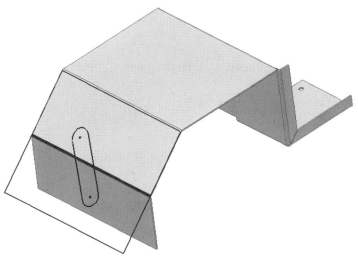

Figure 8.45

17. Pick the Cut tool from the Sheet Metal Panel Bar and click inside the shape. Click the Cut Across Bend option in the Cut dialog box and pick OK to complete the Cut shown in Figure 8.46.

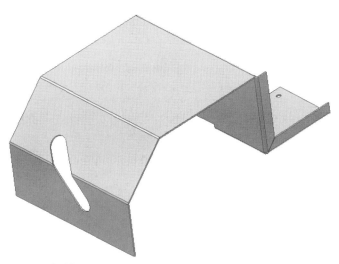

Figure 8.46

18. Create a flat pattern of the folded part by selecting the face highlighted in Figure 8.47 and picking the Flat Pattern tool from the Sheet Metal Panel Bar. The flat pattern model is displayed in a new window as shown in Figure 8.48.

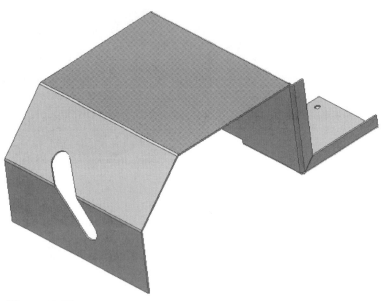

Figure 8.47

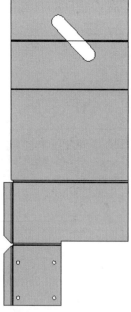

Figure 8.48

19. Export the flat pattern in DXF format by right clicking on Flat Pattern in the Browser and picking Save Copy As from the context menu. Pick DXF Files (*.dxf) from the Save as type list in the Save dialog box and enter **Flat Pattern** for the filename, then click the Options button. Pick the R12 File Version and then click Next. Examine the Style to Layer mapping, then click Finish to close the wizard and then Save to save and exit the Save Copy As dialog box.

 Note: Volo View 2 or Volo View Express 2 needs to be installed for DXF operation to work.

20. Close the Flat Pattern window and click the Return tool on the Standard Toolbar to return to the assembly level. Turn on the visibility of the other two parts in the assembly. The assembly should resemble Figure 8.49.

Figure 8.49

21. Add a Note to the sheet metal part by right clicking the sheet metal part in the graphics window and picking Create Note from the context menu. In the Engineers Notebook, enter **Material to be 1/8" Gold** in the comment box, then right click in the view below the comment box and pick Freeze from the context menu as shown in Figure 8.50. Close the Engineers Notebook window.

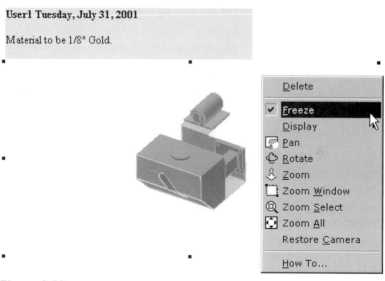

Figure 8.50

22. Finally, make a change to the Sheet Metal Style to reflect the requested material by right clicking the sheet metal part in the graphics window and selecting Edit from the context menu. Pick the Styles tool from the Sheet Metal Panel Bar, and then pick Gold from the Material list. Enter **0.125** in the Thickness edit box, then click on the Bend tab and enter **Thickness * 2** in the Radius edit box. Click Save and then Done. The sheet metal part updates to reflect the changes to the current Sheet Metal Style as shown in Figure 8.51.

23. Return to the assembly and close the assembly file without saving changes.

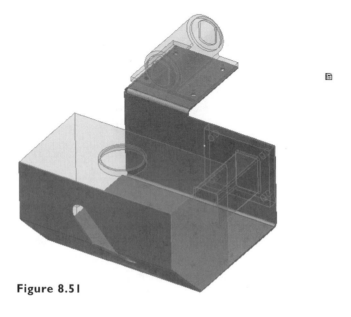

Figure 8.51

PARTS LIBRARY

Autodesk Inventor 5 comes with a library of standard parts including a wide range of fasteners, bearings, seals, and other common mechanical parts. Parts are available in ANSI, ISO, DIN, and other standard formats. The parts library is included on the installation CD. Parts generated from the library are static solids (not parametric) and cannot be edited when saved as Inventor parts. You can optionally generate a SAT file of the part that can be edited using Inventor's solids editing tools.

REDSPARK AND i-DROP

RedSpark supplies the parts library included on the Inventor CD. Additional parts will be available in the future through RedSpark and through Autodesk's PointA web site. The placement of parts from the library is accomplished via a simple browser based interface, and uses Autodesk's i-drop technology to enable drag-and-drop insertion of components in an assembly. The i-drop plug-in installed with the parts library, enables you to drag-and-drop an i-drop enabled Inventor component from any web page into Inventor's graphics window.

LIBRARY SETUP

If there are multiple Inventor users in your design team, the parts library should be installed on a server where all users can access a common set of standard parts. A stand alone installation of Inventor will have faster access to the library if you install it on the same computer. After installing the library and the Inventor plug-in to enable access to the library, you complete the setup by modifying your project file(s), and adding a link to the library index.

Project File: The library stores the parts definitions, not the model itself. The first time a part is selected from the library, an Inventor part file is created and stored in a folder named StandardParts. You can specify the location of this folder prior to selecting parts from the library by adding a library search path named StandardParts to your project file(s). If no such path exists, a folder with this name is added below the current workspace, and a library search path pointing to the folder is added to the current project file. The StandardParts folder is always searched before a library part is created. If the part exists in the library, that part is used.

Shared Content Link: The Place Content tool (see below) enables you to link to an external component source such as the RedSpark parts library. You must add a link to the external source, typically an HTML file or an EXE file, on the Assembly tab of the Application Options dialog box. To link to the parts library, specify the index.htm file in the RedSpark installation folder.

Properties: Library parts contain attributes that are mapped to Inventor properties when the part is generated. Properties such as a manufacturers part number, description, and mass may be assigned values. Select Tools > RedSpark > Catalog Options to edit the mapping between source attributes and Inventor properties.

PLACE CONTENT

Library parts are placed in an assembly using the Place Content tool. To place a library part in an assembly, follow these guidelines.

- Pick the Place Content tool from Assembly Panel Bar as shown in Figure 8.52. Internet Explorer starts and displays the index for the parts library as shown in Figure 8.53.

Figure 8.52

- Browse to the desired library part using the navigation pane on the left side of the page. Optionally, click the Search tab and enter a text search string.

- A preview and options for the selected part are displayed on the right side of the page. Refine the part definition by selecting values from the parameter lists below the part. For a bolt, you would specify thread and length parameters.

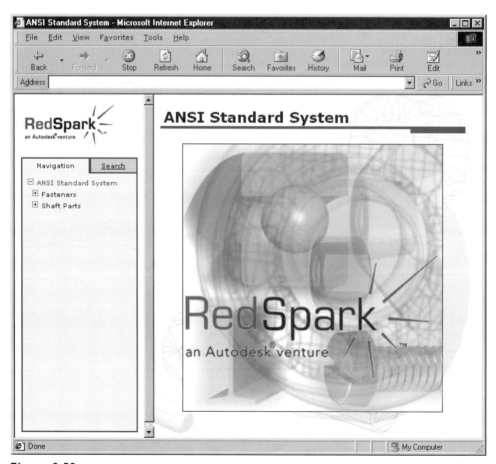

Figure 8.53

- Place the part in the assembly using one of the following techniques:

    i-drop: Click the picture of the part and hold down the mouse button until the eye dropper symbol fills. Drag and drop the part into Inventor's graphics window.

     Note: If the Inventor window is not visible, drag over the Autodesk Inventor icon in the task bar. Pause over the icon for a moment and the Inventor window will activate.

    CAD Button: Click the CAD button below the part picture. You can then choose to directly insert the part in the current assembly in an IPT format, or download the part to a folder in SAT format.

- Constrain the part to the assembly using standard assembly techniques.

TUTORIAL 8.2—FASTENERS AND REDSPARK PARTS LIBRARY

In this tutorial you will place fasteners and RedSpark parts into existing assemblies.

1. If the Fastener library is not installed, install it now from the Inventor CD.
2. Open the assembly \Inventor Book\Linkage.iam that was created earlier in the book and practice placing different fasteners and parts from the RedSpark parts library. When done, close the file without saving the changes.
3. Open the assembly \Inventor Book\Reel.iam that was created earlier in the book and practice placing different fasteners and parts from the RedSpark parts library. When done, close the file without saving the changes.

DESIGN ASSISTANT

Design Assistant is a tool for managing Inventor files and the relationships between files. You can use Design Assistant to view Inventor models, find related files, change or copy file properties using a spreadsheet interface, create a copy of an assembly and its components, and otherwise maintain Inventor files and related documents such as spreadsheets and text files.

You can open Design Assistant from Inventor or from Windows Explorer. Design Assistant can also be installed as a stand alone application on a computer that does not have Inventor installed. This enables other design team members to view and manage Inventor files, properties, and relationships. To start Design Assistant, follow these guidelines.

- **From Inside Inventor:** While a part, assembly, drawing, or presentation file is open, pick File > Design Assistant from the drop-down menu.
- **From Windows Explorer:** Right-click an Inventor file and pick Design Assistant from the pop-up menu.

DESIGN ASSISTANT MODES

Design Assistant has three modes: properties, manage, and preview. The manage mode, which is used to manage relationships between files, is only available if you start Design Assistant from Windows Explorer.

Properties Mode

All Inventor files contain properties that can be used to associate non-graphical data to your models or drawings. A property value can be linked to a field in a drawing title block, or displayed in an assembly bill of materials or drawing parts list. In Properties mode, Design Assistant displays a file hierarchy browser in the left pane, and a spreadsheet view of file properties in the right pane as shown in Figure 8.54. Properties of all first-level children of the current selection in the Browser are displayed in the properties pane. As an example, selecting View Design Properties in the Browser displays the properties of the top-level document. Select an assembly file in the Browser, and properties of the first level children are displayed. The following is a list of tools available from the pull-down menus in Properties mode.

View > Refresh: Reloads all files and properties from their storage location.

View > Customize: Enables you to add and remove property columns from the right pane.

Tools > Reports: Creates and saves a text file containing either the hierarchy of the selected file, or the displayed properties of the selected file. You can expand the coverage to include any number of levels below the selected document.

Tools > Copy Design Properties: Enables you to copy properties between documents. You can select the properties to copy, the file to copy from, and the destination files.

Tools > Find > Autodesk Inventor Files: A criteria based search engine that finds files based on name, type, and properties. The search results can be saved to a text file.

Tools > Find > Where Used: Searches for other files that reference a selected file. You can define search paths (paths in the current project file are the default search paths) and types of files to include in a search. For example, you could search for all drawing and presentation files that reference a selected assembly file.

Tools > Viewer: Opens the viewer that is associated with your computer. Included with the Inventor installation CD is a copy of Volo View Express that can be used to view Inventor files. If another application is registered on your system to view Inventor files, it will be opened. If no viewer is associated with Inventor files, Volo View Express will be registered as the default viewer.

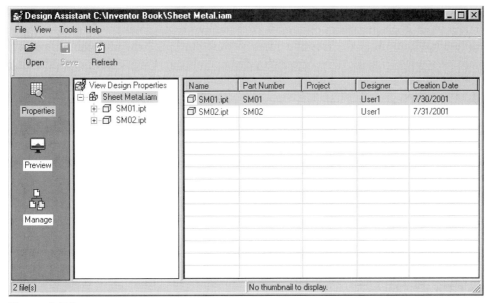

Figure 8.54

Preview Mode

In Preview mode, thumbnail views of selected files are shown in the right pane. By default, a thumbnail view of the active window is included with the file when it is saved. You can turn off the saving of a preview, or specify a BMP file to be used as the preview. You can access the preview settings on the Save tab of the file's Properties dialog box. Files saved without a preview will show a blank preview in Design Assistant.

Manage Mode

Design Assistant's Mange mode is used to manage the structure and relationships of referenced files. You can rename, copy, replace files, and search for files that reference a selected file. The Manage mode window is shown in Figure 8.55.

Manage mode is often used to manage assembly files. All components in an assembly are saved in separate part or subassembly files, and referenced into the assembly. The assembly maintains a list of component names and locations, and loads the referenced file on opening. Renaming a part file outside Design Assistant will result in the assembly being unable to locate the file when it is opened next. Renaming files within Design Assistant updates the information in the assembly, and optionally in other files that reference the renamed file.

You can also copy parts or complete assembly hierarchies (for managing revisions) from within Design Assistant. Assembly components can be replaced from Design

Assistant, without the need for Inventor to be installed on the computer. Care must be taken when replacing components in an assembly; replacement with a dissimilar component will result in the loss of assembly constraints, and the replacement component may not be positioned correctly in the assembly. iParts, or components with matching iMates are good candidates for replacement through Design Assistant.

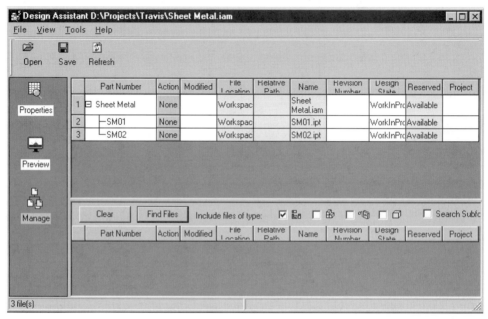

Figure 8.55

DWG (AUTOCAD AND MECHANICAL DESKTOP) WIZARDS

OVERVIEW

Many Inventor users have a large amount of legacy data in AutoCAD or Mechanical Desktop "DWG" format. Other users may require Inventor drawing files be exported in DWG or DXF format to support suppliers or existing manufacturing methods. You can import 2-D drawing data into Inventor drawings, or directly into a part feature sketch. Mechanical Desktop models can be translated into native Inventor files, or linked into an Inventor assembly. Inventor drawings and flat pattern models can be exported in DWG and DXF formats. Import and Export Wizards guide you through the process of translating data between the various formats.

Note: Volo View Express 2 must be installed on the computer for the AutoCAD to Inventor translation to work. Mechanical Desktop 4 or later must be loaded on the computer for the Mechanical Desktop to Inventor translation to work.

IMPORT

AutoCAD and AutoCAD Mechanical Files

Legacy 2-D data can be imported into an Inventor drawing view, or directly into the sketch environment in a part model. The most versatile method for importing "DWG" data is to open the drawing file. The DWG import wizard guides the import process and enables the imported geometry to be saved in a number of formats. Additionally, you can import a "DWG" file into the sketch environment in both the part and drawing environments. Opening a 2-D DWG file directly enables you to save the data to the following destinations:

New Drawing: Creates a new Inventor drawing file and imports geometry into a draft view. Dimensions and annotations are imported; blocks are translated to sketched symbols in the Drawing Resources folder. Dimensions can be promoted to the sketch by right clicking on them and picking Promote to Sketch. The geometry can be copied to the clipboard and pasted onto a sketch in a part "ipt" file.

New Part: Creates a new part file and imports geometry to Inventor sketch geometry. Dimensions, text, and other annotations are not imported.

Title Block: Creates a new Inventor drawing file and imports geometry, including annotations, into a new title block.

Border: Creates a new Inventor drawing file and imports geometry, including annotations, into a new border.

Symbol: Creates a new Inventor drawing file and imports geometry, including annotations, into a new sketched symbol. The sketched symbol is stored in the drawing and an occurrence of the symbol is placed on the drawing sheet.

To import 2-D geometry into the active sketch (in a part or drawing), pick the Insert AutoCAD File icon from the Sketch Panel Bar as shown in Figure 8.56. The DWG Import Wizard guides you through the steps required to import the 2-D geometry into the sketch. See the import options above for information on the types of entities imported to the specific sketch environment.

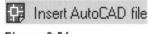
Figure 8.56

You can also import AutoCAD files containing 3-D solids. When you import an AutoCAD file with 3-D solids, each solid is translated to an Inventor part file. An Inventor assembly file is created and all translated parts are inserted into the assembly. Parts are placed in the same position and orientation as the AutoCAD file and given the same name as the AutoCAD file with a sequenced number at the end of the part name. For example, an AutoCAD file named "Piston.dwg" that contained two solids would be imported into an Inventor assembly named "Piston.iam" and the parts would be named "Piston1.ipt" and "Piston2.ipt".

Import Wizard

The DWG Import Wizard is activated automatically when opening a DWG file. You can also access the wizard by picking the Options button in the Open dialog box when a DWG file is selected. To import an AutoCAD file into a new Inventor file, follow these guidelines.

- Pick File > Open from the pull-down menu.
- In the Open dialog box, pick AutoCAD Drawing (*.dwg) from the Files of type list.
- Browse to and pick the desired DWG file.
- Click Open to start the DWG File Import Options wizard as shown in Figure 8.57. The type of DWG file, AutoCAD, AutoCAD Mechanical, or Mechanical Desktop, is pre-selected.

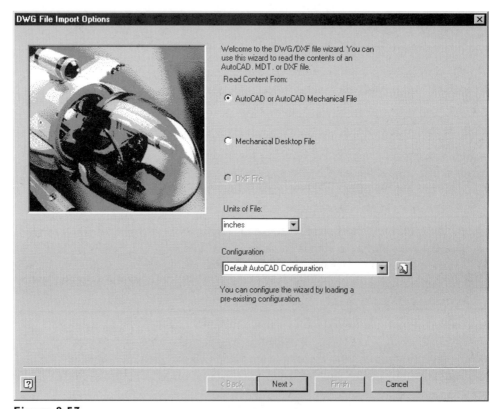

Figure 8.57

- Pick the units used in the DWG file from the Units of File list.
- Optionally, pick a saved configuration to use translation options from a previous import.
- Click Next. The Layer and Objects Import Options dialog box is displayed as shown in Figure 8.58.

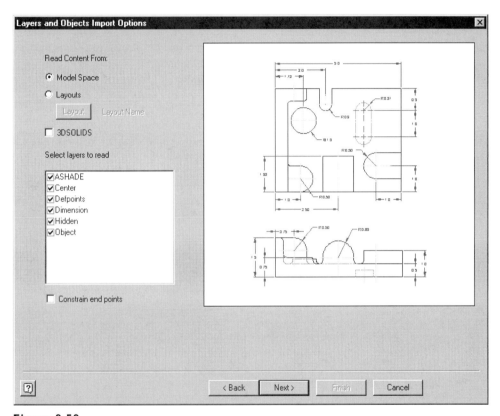

Figure 8.58

- Pick to Import Objects from either Model Space or from a Layout within the DWG file.
- Choose which layers you want to import.
- Optionally, pick to Constrain End Points. This option verifies that the adjoining endpoints of imported geometry are constrained.
- Click Next. The Import Destination Options dialog box is displayed as shown in Figure 8.59.

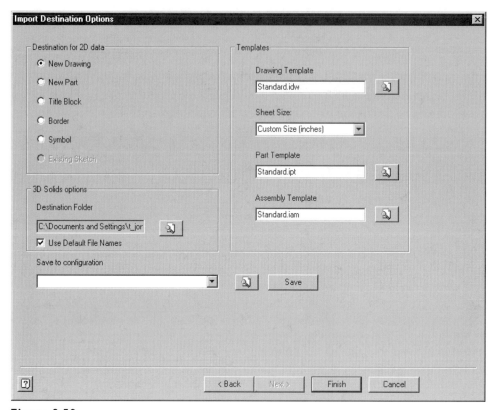

Figure 8.59

- Select the Destination for the imported data. You can import 2-D data into a draft view in a new drawing or as the contents of a sketch in a new part file. The data can also be imported into a title block, border, symbol, or an existing sketch.
- Alternatively, you can select to import 3-D solids as described above.
- When importing 3-D solids, you can check the Use Default File Names box to speed up the process of importing data. If the box is unchecked, you will be prompted to specify part names for each 3-D solid. 3-D Solids can be imported from either model space or paper space, but not both simultaneously.
- Click Finish to import the data.

Mechanical Desktop Files

Mechanical Desktop part and assembly models can be translated into native Inventor files that retain part feature history and assembly constraints. You can also link Mechanical Desktop parts in an Inventor assembly without translation. The part appears as a native Inventor part (ipt file), but cannot be edited in Inventor. You can edit the original part in Mechanical Desktop and update the linked part in the Inventor assembly. Drawing views in Mechanical Desktop files can be imported into a new Inventor drawing (idw file). The Inventor model and drawing files are not associative; changes to the imported model will not be reflected in the imported drawing.

Inventor has different types of files than Mechanical Desktop. Each Inventor file (parts, assemblies, and drawings) is saved in its own file. Mechanical Desktop can combine all three within the same file. When importing Mechanical Desktop files, the DWGIN Wizard will present options specific for this difference. The first page of the wizard, DWG File Import Options, works the same way it's described above. In the Import Wizard dialog box clicking Next will open the MDT Model/Layout Import options page as shown in Figure 8.60.

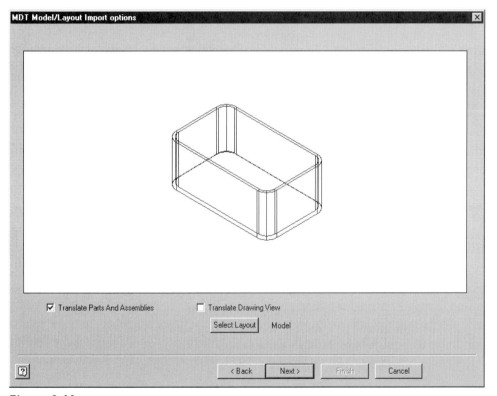

Figure 8.60

- Select the type of object(s) to translate—Parts and Assemblies, Drawing Views, or both. You can select a specific Layout from the DWG file if multiple Layouts are present in the DWG file.

- Click Next. The MDT Import Destination Options dialog box is displayed as shown in Figure 8.61.

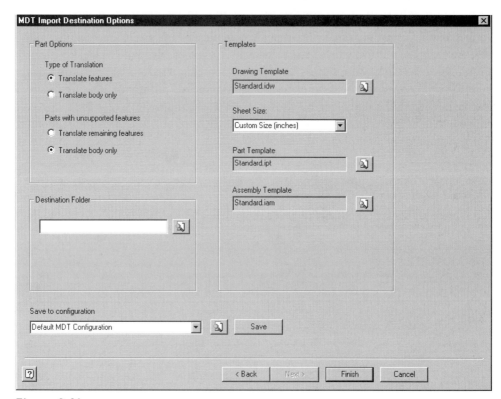

Figure 8.61

- Select to translate all features into similar Inventor features or to translate the MDT file into an Inventor static solid. During the import operation, you may run into an error where Inventor cannot translate a particular feature. Then, through a dialog box, you can select how you want Inventor to continue with the translation process if an error occurs.
- Specify a destination folder.
- Choose the type of drawing, part, or assembly template, and the sheet size for the drawing.
- Optionally, choose to save the settings as a configuration to use the translation options in a later import.
- Click Finish to import the data.

EXPORT

Inventor drawing files can be exported to various DXF and DWG versions. Each sheet in the file is exported to a separate DWG file. Drawing lines and curves are exported to model space at a 1:1 scale, and lines, curves and all drawing annotations are exported to a layout in the DWG file. The DXF/DWG File Export Options wizard guides you through the configuration of the translation. Translation settings can be saved in a translation configuration that can be recalled during subsequent exports. The wizard enables you to map Inventor drawing entities to DWG layers, specify a template file to be used when generating the DWG/DXF file, and gives you control over the export of parts list information to an AutoCAD Mechanical file. The exported DWG or DXF file is not associative to the Inventor drawing file. To export an Inventor drawing in DWG format, follow these guidelines.

 Note: The options available when exporting to a DXF format are a subset of the DWG export options.

- Pick File > Save Copy As from the drop-down menu.
- In the Save dialog box, pick AutoCAD Drawing (*.dwg) from the Save type list.
- Click the Options button to start the DWG File Export Options wizard as shown in Figure 8.62.

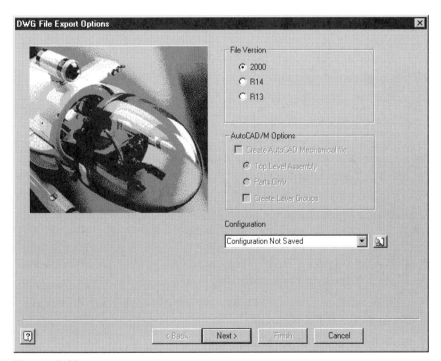

Figure 8.62

- Pick the AutoCAD file version.
- Pick the AutoCAD\M Options if saving to that format.
- Optionally, pick a saved configuration to use translation options from a previous export.
- Click Next.
- Optionally, if the export is to an AutoCAD Mechanical drawing, the Parts List Export Options dialog box will be displayed. Add the properties to be exported to the AutoCAD Mechanical parts list. Click Next.
- The Layers Export Mapping dialog box is displayed as shown in Figure 8.63.

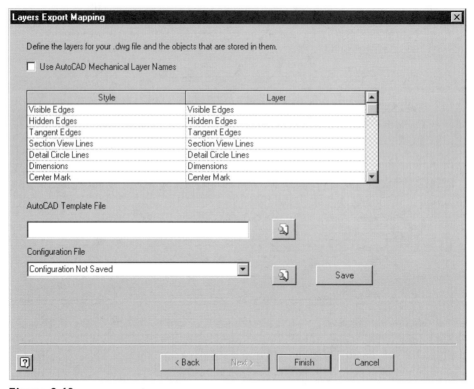

Figure 8.63

- Specify whether to use the AutoCAD Mechanical layer names.
- Edit any of the layer names associated with an Inventor style.
- Optionally, select an AutoCAD template file that will be used as the basis for the exported DWG file.
- Click Finish.

 Note: You can save the settings in a configuration file to be used in future file exports. Save a different configuration for each common file export.

TUTORIAL 8.3—DWG IMPORT

In this tutorial you will import a DWG file into a new Inventor drawing file, promote the dimensions to parametric dimensions, and then copy the geometry and dimensions into a part's sketch. You will complete the tutorial by starting a new part file and following the steps to bring an AutoCAD 2-D file directly into a part's sketch.

1. Start a new drawing file based on the default Standard.idw file.
2. Pick the Sketch tool from the Standard toolbar. The sketch environment and tools will become active.
3. Pick the Insert AutoCAD file tool from the Sketch Panel Bar.
4. In the Open dialog box select the file \Inventor Book\AutoCAD 2D Exercise.dwg.
5. Pick the Options button, then pick the Next button from the DWG File Import Options dialog box to accept the defaults on the first page.

   In the Layers and Objects Import Options dialog box uncheck the Border and Defpoints layers and check the Constrain end point box as shown in Figure 8.64.

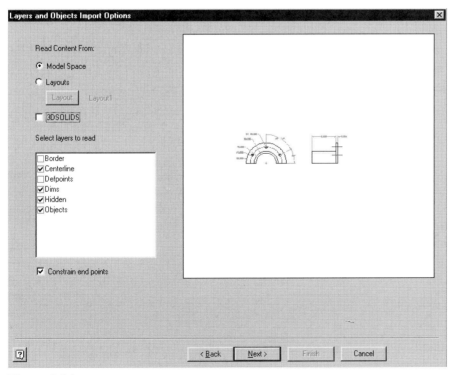

Figure 8.64

6. Pick the Next button and then from the Import Destination Options select the Finish button to accept the defaults.
7. In the Open dialog box select Open.
8. The geometry and dimensions will be placed in the sketch.
9. Select the dimensions, right click, and pick Promote to Sketch.
10. Change the 3.000 dimension in the right side view to 4.000 and then back to 3.000.
11. Select the geometry in the front view and copy it to the clipboard by right clicking and pick Copy from the Context menu.
12. Start a new part file based on the default Standard.ipt file.
13. Paste the objects from the clipboard by right clicking and picking Paste from the context menu.
14. Change a few of the dimensions to verify that they are parametric.

    Using the shared sketch technique turn the sketch into a parametric part. Extrude both areas .25 to the back, share the sketch, and then extrude just the inside area 3 inches. When done, your part should resemble Figure 8.65.

Figure 8.65

15. Close both the drawing and part file without saving the changes.
16. Start a new part file based on the default Standard.ipt file.
17. Pick the Insert AutoCAD file tool from the Sketch Panel Bar.
18. In the Open dialog box select the file \Inventor Book\AutoCAD 2D Exercise.dwg
19. Pick the Options button, then pick the Next button from the DWG File Import Options dialog box to accept the defaults on the first page.
20. In the Layers and Objects Import Options dialog box uncheck every layer except Objects and Border. Check the Constrain end point box.

     Note: Dimensions will not import directly into a sketch.

21. Pick the Next button and then from the Import Destination Options select the Finish button to accept the defaults.
22. In the Open dialog box select Open.

23. The geometry will be placed in the sketch.
24. Add dimensions to fully constrain the sketch. Use your own values for the dimensions.
25. Using the shared sketch technique, turn the sketch into a parametric part.
26. Close both the drawing and part file without saving the changes.
27. Practice importing your own AutoCAD files into both Inventor drawings and parts.

IMPORTING OTHER FILE TYPES

Autodesk Inventor can import parts and assemblies exported from other CAD systems. Inventor models created from these formats are base solids or surface models, no feature histories or assembly constraints are generated when importing a file in one of the following formats. You can add features to imported parts, edit the base solids using Inventor's solids editing tools, and add assembly constraints to the imported components.

SAT: A neutral file format available from most CAD systems based on the ACIS modeling kernel (Autodesk Inventor, AutoCAD, and Mechanical Desktop use this modeling kernel). CAD systems based on other kernels can often export parts and assemblies in this format. SAT files can be imported either as solid bodies, or as surface bodies using the Insert > Import pull-down menu option in the part environment.

STEP: The STEP format is widely used to translate 3-D solid models between CAD systems. Inventor can import models saved in AP-203 or AP-214 STEP formats.

Pro/E: Inventor can automatically translate and import native Pro/E files up to version 20. Complex solid models and Pro/E surfaces can cause import problems. Pro/E files exported in STEP format are more reliably imported. Pro/E 2000i and newer files must be exported to another format for import into Inventor.

IGES: Early data translation between CAD systems almost always used the IGES file format, and it is still commonly used today. IGES files can contain many different types of data, but typically consist of a set of surfaces representing the boundary of the part. An IGES file is imported into a Construction folder as a Group of surfaces. You can then select surfaces and stitch them into a surface quilt. If the surfaces can be combined into a surface quilt with no gaps, the quilt can be promoted to an Inventor base solid. You can promote individual surfaces and surface quilts to grounded Inventor surfaces. These surfaces cannot be edited in Inventor. Surface healing of IGES data is not done. Gaps between surfaces due to file precision or other factors will prevent an imported IGES file from being promoted to a base solid.

REVIEW QUESTIONS

1. An Engineering Note can be attached to a drawing file. True or False?
2. Sheet Metal Style settings cannot be overridden, a new Style must be created for different settings. True or False?
3. During the creation of a sheet metal face, it can extend to meet another face and connect to it with a bend. True or False?
4. You can add standard part features to a sheet metal part. True or False?
5. A sheet metal flat pattern is a 2-D representation of the unfolded model. True or False?
6. You can use Design Assistant to copy file properties between unrelated files. True or False?
7. With Design Assistant, you can copy parts, but not assemblies. True or False?
8. You must be in the sketch environment to import a 2-D AutoCAD file. True or False?
9. Inventor layers are translated to AutoCAD layers during export of an IDW file. True or False?
10. Imported SAT files retain the creation history of the part. True or False?

INDEX

A

Active component, 194
Adaptivity, 226–230
    tutorial, 230–233
Alternate Title Block Alignment, 141
Alt Units, 151
Angle assembly constraint, 202, 204, 206
Animation, 240–242
Annotations, 175–179
    centerlines, 175–177
    surface textures, weld symbols, and feature control frames, 177
    tutorial, 182–185
ANSI, 148–150, 179
Application options, 7–14
    colors tab, 10
    display tab, 11–13
    file tab, 9
    general tab, 7–8
    hardware tab, 14
Apply Driven Dimensions, 20
Assemblies, 188–255
    adaptivity, 226–230
    adaptivity tutorial, 230–233
    bottom up approach, 192
    bottom up tutorial, 196–197, 212–214
    constraints, 201–208
    creating, 189–190
    creating drawing balloons, 246–248
    creating drawing views from assemblies and presentation files, 245–246
    design views, 234–235
    drawing views tutorial, 249–252
    driving constraints, 209–211
    editing assembly constraints, 208–209
    enabled components, 234
    grounded components, 194–195
    iMates, 218–220
    iMates, replacing, and pattern components tutorial, 222–226
    In-Part features, 191–192
    interference checking, 211–212
    Part Feature Adaptivity, 191
    parts list, 248–249
    patterned components, 220–222
    presentation files, 235–242
    presentation views tutorial, 242–245
    reorder and restructure, 196
    replacing components, 220
    Shared Content Link, 192
    subassemblies, 195–196
    top down approach, 193–194
    top down tutorial, 197–201, 215–218
    work features, 226
Assembly files, 6
Auto baseline dimensions, 168–169
Auto Dimension, 28, 270–271
    tutorial, 277–282
Auto Stitch and Promote, 22
Auto-Bend with 3-D Line Creation, 21
Auto-Hide In-Line Work Features, 21
AutoCAD
    2-D drawing creation, 181
    exporting and, 395–396
    importing and, 389–392, 397–399
Autodesk Inventor Browser, 50–51
Automatic constraints, 30
Automatic Reference Edges for New Sketch, 21

AutoProject Edges During Curve Creation, 21
Auxiliary views, 152, 155
Axes, 20
Axis, 59

B

Background Gradient, 10
Balloons, 149
   creating, 246–248
Base features, 50
Base views, 151, 152–154
Bend tool, 364
Blending Transparency, 12
Borders
   creating, 143
   tutorial, 145–148
Bottom up assembly approach, 192
   tutorial, 196–197, 212–214
Broken views, 152, 157–158
Browser, 16, 50–51
Browser Toolbar, 16
BSI, 148–149, 179

C

Camera views, 45
Catalog Tools, 370
Center Point Arc, 27
Center Point Circle, 27
Centerlines, 60–63, 175–177
Centermark, 149
Chamfers, 27, 52, 96–101
   tutorial, 98–101
Circular patterns, 28, 53, 132
   tutorial, 134–136
Coil, 52
Coil features, 300–303
   tutorial, 303–308
Coincident constraint, 33
Colinear constraint, 33
Color Schemes, 10

Color tab, 10
Command Bar, 15
Command Entry, 17–19
Common, 149
Common View (Glass Box), 47–48
Components
   enabled, 234
   iMates, replacing, and pattern components tutorial, 222–226
   patterned, 220–222
   replacing, 220
   tweak, 238–240
Concentric constraint, 33
Constant tab, 90–91
Constraint Placement Priority, 19
Constraints, 28
   adding, 35–36
   assembly, 201–211
   over-constrained sketches, 38
   showing and deleting, 34–35
   sketching, 32–36
   symmetry, 260
   tutorial, 39–41
   types of, 33–34
Construction geometry, 257
Contour flange tool, 359–360
Control Frame, 149
Coordinate system indicator, 20
Copy/paste features, 316–317
   tutorial, 317–321
Corner chamfer tool, 367–368
Corner round tool, 367
Corner seam tool, 364–365
Create iFeature, 53
Creating a part. *See* Part creation
Cut, 73–74
Cut tool, 366

D

Datum Target, 149
Default title block, 143–144

Index

Default VBA Project, 9
Defer Update, 8
Deleting drawing views, 166
Depth Dimming, 11, 12
Derived assembly, 343–344
   tutorial, 346–347
Derived Component, 53
Derived parts, 340–344
   tutorial, 344–346
Design Assistant, 385–388
   manage mode, 387–388
   preview mode, 387
   properties mode, 386
Design Support System, 14–15
Design views, 234–235
Detail views, 152, 156–157
Diametric dimensions, 37, 60–63
Dimension display, 271–272
Dimension relationships, 272
Dimension styles, 149–151
Dimensioning
   angles, 37
   arcs and circles, 37
   entering and editing values, 38
   lines, 37
   to a quadrant, 38
   tutorial, 39–41
Dimensions, 166–175
   adding, 36–38
   auto baseline, 168–169
   dimension visibility, 166–167
   drawing (reference), 167
   hole tables, 172–173
   moving, 172
   ordinate, 169–172
   tutorial, 173–175
   value and appearance of, 167
DIN, 148–149, 179
Direction, 56
   for shelling, 113
Display, 19, 151
Display options, viewing, 48–49
Display Quality, 11
Display tab, 11–13
Distance, 56
Distance and Angle, for chamfers, 97–98
Draft views, 158
Drafting standards, 148–149
Drag and drop components, 194
Drawing, creating, 141
Drawing files, 7
Drawing options, 140–141
Drawing (reference) dimensions, 167
Drawing sheets preparation, 142–143
Drawing tools, 142
Drawing views, 139–187
   annotations, 175–177
   annotations tutorial, 182–185
   assembly tutorial, 249–252
   auto baseline dimensions, 168–169
   auxiliary views, 155
   base views, 152–154
   border creation, 143
   broken views, 157–158
   creating an AutoCad 2-D drawing from Inventor drawing views, 181
   creating a drawing, 141
   creating from assemblies and presentation files, 245–246
   deleting, 166
   detail views, 156–157
   dimension styles, 149–151
   dimension visibility, 166–167
   dimensions value and appearance, 137
   draft views, 158
   drafting standards, 148–149
   drawing options, 140–141
   drawing (reference) dimensions, 167
   drawing sheets preparation, 142–143
   drawing tools, 142
   edit drawing view properties, 164–166

hole tables, 172–173
hole and thread notes, 178
moving, 164
moving dimensions, 172
ordinate dimensions, 169–172
parts lists, 178–181
projected views, 154
section views, 155–156
sketched symbols, 177–178
template, 151
title block creation, 143–145
tutorial, 145–148, 159–164, 173–175
Drive Constraint, 209–211
DWG (AutoCAD and Mechanical Desktop) wizards, 388–396
Export, 395–396
Import, 389–394
Import tutorial, 397–399
overview, 388
Dynamic Rotate, 46

E

Edge Chain and Setback, for chamfers, 98
Edge Display, 12
Edge and Face, for chamfers, 98
Edit Coordinate System, 27
Edit Dimensions When Created, 21
Edit drawing view properties, 164–166
Ellipses, 27, 257–258
Enabled components, 234
Engineers Notebook, 350–353
adding notes to a model, 351–353
adding other content, 353
managing notes, 353
note display, 353
Equal constraint, 34
Equations, 272
Export, 395–396
Extend, 28
Extents
extruding the sketch, 55–56
revolving the sketch, 60
Extruding, 52
extruding a sketch tutorial, 56–58
the sketch, 53–58

F

F4 Shortcut Dynamic Rotation, 48
Face cycling, 71–72
Face draft, 52, 91, 115–118
draft angle, 116
faces, 115
Pull Direction, 115
tutorial, 116–118
Face tool, 357–359
Failed features, 65
Feature control frames, 177
Features, 50–53, 229
Coil, 300–303
Coil tutorial, 303–308
and color, 66
copy/paste, 316–317
copy/paste tutorial, 317–321
deleting, 66
editing, 64–69
editing a feature's sketch tutorial, 68–69
editing tutorial, 67–68
failed, 65
iFeatures, 332–334
iFeatures tutorials, 334–340
Loft, 308–310
Loft tutorial, 310–312
Mirror, 322–323
mirroring about a plane tutorial, 323–324
renaming, 65–66
reorder, 322
sketched features, 70

suppressing, 325
switching environments, 51
tools for, 52–53
File properties, 347–349
File tab, 9
Files
included, 3
presentation, 235–242, 245–246
projects, 5–6
types of, 6–7
Fillets, 27, 52, 89–96
Constant tab, 90–91
Setback tab, 92–93
tutorial, 93–96
Variable tab, 91–92
Fix constraint, 34
Flange tool, 361
Flat pattern tool, 369–370
Fold tool, 363

G

GB, 148–149, 179
General dimensioning, 28, 36
General tab, 7–8
Geometry, construction, 257
Get Model Dimensions on View Placement, 140
Glass Box (Common View), 47–48
Graphics Window, 16
Grid lines, 19
Grounded components, 194–195

H

Hardware tab, 14
Hatch, 149
Hem tool, 361–362
Hidden edge display, 49
Hidden Edges, 11–12
Highlight Invalid Annotations, 141

Hole notes, 178
Hole tables, 172–173
Holes, 52, 101–107, 370
hole centers, 104–105
Options tab, 104
Size tab, 103–104
Threads tab, 103
tutorial, 105–107
Type tab, 102–103
Horizontal constraint, 19, 34
Hot keys, 18–19

I

i-Drop, 383, 385
iFeatures, 332–334
tutorials, 334–340
IGES file format, 399
iMates, 218–220
Importing, 389–394, 399
DWG tutorial, 397–399
In-place features, 191–192
Included file, 3
Inferred points, 30
Insert assembly constraint, 202, 204, 207
Insert AutoCAD, 29
Insert iFeature, 53
Interference checking, 211–212
Intersect, 73–74
iParts, 282–291
creating, 283–286
placement, 286–288
tutorial, 288–291
ISO, 148–149, 179
Isometric view, 45

J

JIS, 148–149, 179
Join, 73–74

L

Library search paths, 4
Line, 27
Line tool, 29–30
Local search paths, 4
Locate Tolerance, 8
Location, 5
Location tab, for threads, 108–109
Loft, 52
Loft features, 308–310
   tutorial, 310–312
Look At, 47

M

Mate assembly constraint, 202, 204–206
Maximum Size of Undo File (Mb), 7
Mechanical Desktop files, importing
   and, 392–394
Minimum Frame Rate (Hz), 13
Minor grid lines, 20
Mirror, 28
Mirror features, 53, 322–323
   tutorial, 323–324
Mirror and Features Patterns, 370
Mirror sketches, 260
   tutorial, 261–263
Move, 28
Moving drawing views, 164
Multi User, 8
Multiple documents, 194

N

No New Sketch, 21
Number of Versions to Keep, 7

O

Occurrences, 193
Offset, 28
Open, 2
Open and edit, 194
Operation
   extruding the sketch, 54–55
   revolving the sketch, 59–60
Operations and terminations tutorial,
   75–84
Options, 4–5, 151
   application, 7–14
   assembly, 190–192
   drawing, 140–141
   part, 21
   sketch, 19–21
Options tab, for holes, 104
Ordinate dimensions, 169–172
Overconstrained dimensions, 20–21

P

Pan View, 46
Panel Bar, 16–17
Parallel constraint, 19, 33
Parallel View on Sketch Creation, 21
Parameters, 272–275
   linked to a spreadsheet, 275–282
Part creation, 24–25
   new sketch, 25
   sketches and default planes, 24–25
Part feature adaptivity, 191
Part files, 6
Part options, 21
Part Split, 312–313
   tutorial, 314–315
Parts
   derived, 340–344
   derived tutorial, 344–346
Parts library, 383–385
   library setup, 383
   place content, 384–385
   Redspark and i-Drop, 383–384
Parts lists, 149, 178–181, 248–249
Paste. *See* Copy/paste features

Pattern sketches, 258–259
Patterned components, 220–222
Patterns, 130–136
    circular, 28, 53, 132, 134–136
    rectangular, 28, 53, 130–131, 133–134
Percentage Hidden Line Dimming, 13
Perpendicular constraint, 19, 33
Place Content tool, 384–385
Placed features, 50, 88–138
    chamfers, 96–101
    creating a work axis, 119–121
    creating work planes, 121–130
    creating work points, 119–121
    face draft, 115–118
    fillets, 89–96
    holes, 101–107
    patterns, 130–136
    shelling, 111–114
    threads, 108–111
Point, Hole Center, 28
Polygon, 28
Precise input, 31
Precise View Generation, 140
Prefix/Suffix, 151
Presentation files, 6–7, 235–242, 245–246
Presentation views, tutorial, 242–245
Previous View, 48
Pro/E file format, 399
Profile, 54, 59
Project Cut Edges, 29
Project edges, 264–265
Project files, 5–6, 7
    creating a project, 5
    editing a project, 5
    tutorial, 6
Project Flat Pattern, 29
Project Folder, 9
Project Geometry, 29
Projected views, 151, 154
Projects, 3–5
Promote, 53, 370
Pull Direction, for face draft, 115
Pull Down Menus, 15
Punchtool, 368–369

R

Rectangular patterns, 28, 53, 130–131
    tutorial, 133–134
Redo, 19
Redspark, 383–384
    tutorial, 385
Remove Faces, for shelling, 112
Reorder features, 322
Reorder and restructure, 196
Replacing components, 220
Revolving, 52
    the sketch, 59–60
    sketch tutorial, 61–63
Rib, 52
Ribs and webs, 326–328
    tutorial, 328–332
Rotate, 28
Rotation motion constraint, 202, 205
Rotation-Translation motion constraint, 202, 205

S

SAT file format, 399
Save DWF Format, 140
Screen Door Transparency, 12
Scrubbing, 30–31
Section views, 152, 155–156
"Select Other" Time Delay (sec), 8
Setback tab, 92–93
Shaded display, 48
Shaded Display Modes, 12–13
Shape
    extruding the sketch, 54
    revolving the sketch, 59

Shared content link, 192
Shared sketches, 259–260
Sheet, 149
Sheet metal files, 7
Sheet metal parts, 353–354
    creating, 354
    design methods, 354
    tutorial, 370–382
Sheet metal tools, 354–370
    bend, 364
    Catalog Tools, 370
    contour flange, 359–360
    corner chamfer, 367–368
    corner round, 367
    corner seam, 364–365
    cut, 366
    face, 357–359
    flange, 361
    flat pattern, 369–370
    fold, 363
    hem, 361–362
    Holes, 370
    Mirror and Feature Patterns, 370
    Promote, 370
    punchtool, 368–369
    sheet metal styles, 355–357
    Work Features, 370
Sheets, tutorial, 145–148
Shell, 52
Shelling, 111–114
    Direction, 113
    Remove Faces, 112
    Thickness, 112
    tutorial, 113–114
    Unique Face Thickness, 113
Shortcut, 5
Shortcut keys, 17
Shortcut menus, 17
Show 3-D Indicator, 8
Show Constraints, 29
Show Line Weights, 140
Show Reflections and Textures, 10
Show Startup Dialog, 8
Silhouettes, 11–12
Size tab, for holes, 103–104
Sketch on New Part Creation, 21
Sketch options, 19–22
Sketch on X-Y Plane, 21
Sketch on X-Z Plane, 21
Sketch on Y-Z Plane, 21
Sketched features, 50
    See also Sketching
Sketched symbols, 177–178
Sketches, 228–229
    mirror, 260–263
    pattern, 258–259
    shared, 259–260
    See also Sketching
Sketching
    adding constraints, 35–36
    adding dimensions, 36–38
    on another part's face, 265–270
    assigning a plane to the active sketch, 70–73
    automatic constraints, 30
    constraining the sketch, 32–36
    constraint types, 33–34
    and default planes, 24–25
    deleting objects, 32
    dragging, 33
    editing a feature's sketch tutorial, 68–69
    extruding the sketch, 53–58
    extruding tutorial, 56–58
    inferred points, 30
    line tool, 29–30
    new sketch, 25
    outlining the part, 26–32
    over-constrained sketches, 38
    overview, 26
    precise input, 31
    renaming, 65–66

revolving a sketch tutorial, 61–63
scrubbing, 30–31
selecting objects, 31
showing and deleting constraints, 34–35
sketch features, 70
sketch planes tutorial, 72–73
tips for, 26
tools for, 27–29
tutorial, 32
Slice graphics, 263–264
Snap 'n Go constraints, 207–208
Snap to Grid, 21
Specification tab, for threads, 109
Split, 53
Splitting a part. *See* Part Split
Spreadsheet view, 181
Standard Toolbar, 15
Status Bar, 16
STEP file format, 399
Subassemblies, 194–196, 229
Surface textures, 149, 177
Surfaces, using, 326
Sweep, 52, 298–299
  tutorial, 299–300
Symmetry, 34
Symmetry constraint, 260

T

Tangent Arc, 27
Tangent assembly constraint, 202, 204, 206
Tangent Circle, 27
Tangent constraint, 33
Taper, 54
Templates, 9, 23, 151
Termination, 55
Terminations and operations tutorial, 75–84
Terminator, 149, 151

Text, 151, 178
Thickness, for shelling, 112
Thread notes, 178
Threads, 52, 108–111
  Location tab, 108–109
  Specification tab, 109
  tutorial, 109–111
Threads tab, for holes, 103
Three-D Sketch, 21
Three Point Arc, 27
Three Point Rectangle, 27
Title block creation, 143–145
Title blocks, tutorial, 145–148
Tolerance, 151
Toolbars, 16–17
  Browser, 16
  Standard, 15
Top down assembly approach, 193–194
  tutorial, 197–201, 215–218
Transcripting, 9
Trim, 28
Tweak components, 238–240
Two Distances, for chamfers, 98
Two Point Rectangle, 27
Two-D design layout, 292
  tutorial, 292–294
Type tab, for holes, 102–103

U

Undo, 9, 19
Unique Face Thickness, for shelling, 113
Units, 22–23, 151
Use relative paths, 4
User Interface, 15–17
Username, 8

V

Variable tab, 91–92
Vertical constraint, 19, 34

View Catalog, 53
View Transition Time, 13
Viewing, 44–49
   camera views, 45
   display options, 48–49
   isometric, 45
   tools for, 45–48
   tutorial, 49
   See also Drawing views
Views
   design, 234–235
   See also Drawing views

W

Warn of Overconstrained Condition, 20
Webs. See Ribs and webs
Weld symbols, 149, 177
Windows shortcut keys, 17

Wireframe display, 49
Wireframe Display Mode, 11
Work Axis, 53, 119
   tutorial, 120–121
Work Features, 370
Work Planes, 53, 121–123
   tutorials, 123–130
Work Points, 53, 119
   tutorial, 120–121
Workgroup Design Data, 9
Workgroup search paths, 4
Workspace, 4

Z

Zoom All, 46
Zoom Selected, 46
Zoom Window, 46